High Octane Heritage
Celebrating the Oldsmobile 442

Todd Bandel

Copyright © 2024 Todd Bandel

All rights reserved.

ISBN: 9798301172212

DEDICATION

I dedicate this book to all my past automotive mentors and colleagues. Your guidance, support, and shared wisdom have been invaluable in shaping my journey. Each of you played a significant role in my professional development, imparting knowledge and fostering a passion for excellence in the automotive field.

Contents

ACKNOWLEDGEMENT i

CHAPTER ONE
The Birth of a Legend: Origins and Early Years of the Oldsmobile 442 1

Chapter Two
Engineering Marvels: The Heart of the 442 15

Chapter Three
Design Evolution: Styling Changes Through the Decades 31

Chapter Four
Performance Powerhouse: The 442's Engine and Drivetrain 47

Chapter Five
Handling and Suspension: Balancing Power with Control 63

Chapter Six
Special Editions and Rare Variants: Collector's Dream Machines 79

Chapter Seven
The 442 on the Track: Racing Heritage and Achievements 93

Chapter Eight
Cultural Icon: The 442's Impact on American Car Culture 109

Chapter Nine
Restoration Guide: Bringing a 442 Back to Life 125

Chapter Ten
Preserving the Legacy: Restoration Techniques and Challenges 139

Chapter Eleven
Maintenance Mastery: Keeping Your 442 in Prime Condition 155

Chapter Twelve
The 442 Community: Stories from Owners and Enthusiasts 171

ACKNOWLEDGMENTS

I want to express my deepest gratitude to my father for introducing me to the exhilarating world of automotive racing. Your passion for cars and dedication to the sport have inspired me.

From the first time you took me to a race track, I was captivated by the power and precision of the machines, as well as the skill required to master them.

Your guidance and support have fueled my interest and enthusiasm, making every moment in this thrilling world more meaningful. Thank you for sharing this incredible journey and for being such a pivotal influence in my life.

Chapter 1: The Birth of a Legend: Origins and Early Years of the Oldsmobile 442

Section 1.1: The Muscle Car Revolution

The 1960s marked a pivotal era in American automotive history, witnessing the meteoric rise of performance cars that would forever transform the industry's landscape. This period, aptly named the muscle car revolution, set the stage for legendary vehicles that continue to captivate enthusiasts to this day.

As the decade dawned, a shift in consumer preferences began to emerge. The post-war generation, now coming of age, craved excitement and speed. They weren't content with the large, comfortable cruisers that dominated the previous decade. Instead, they yearned for cars that could deliver heart-pounding acceleration and impressive quarter-mile times. This desire for performance sparked a fierce competition among American automakers to produce the fastest, most powerful vehicles they could muster.

Oldsmobile, a division of General Motors, found itself at a crossroads during this transformative period. Known for producing reliable, comfortable cars that appealed to a more mature demographic, Oldsmobile's position in the market was solid but uninspiring to younger buyers. The brand's lineup, while respectable, lacked the excitement and Raw power that was beginning to define the era.

As rivals began introducing high-performance models, Oldsmobile recognized the need for a game-changing entry in its lineup. The success of cars like the Pontiac GTO, which essentially created the muscle car category in 1964, served as both a wake-up call and a blueprint for what was possible. Oldsmobile needed a performance model that could compete with these new titans of the street while maintaining the brand's reputation for quality and innovation.

The automotive climate of the early 1960s was ripe for innovation. Advancements in engine technology, coupled with a booming economy and a young, affluent customer base, created the perfect storm for the muscle car's ascension. Fuel was cheap, safety regulations were still in their infancy, and the American public's appetite for speed seemed insatiable.

It was in this environment of fierce competition and rapid innovation that the Oldsmobile 442 was born. The automotive landscape was shifting dramatically, and Oldsmobile needed to adapt or risk being left behind. The 442 represented more than just a new model; it was Oldsmobile's answer to the changing times, a bold statement that the brand could compete with the best in this new era of high performance.

The muscle car revolution had begun, and Oldsmobile was determined to be at the forefront. The stage was set for the introduction of a car that would not only meet the demands of this new market but exceed them, cementing its place in automotive history and the hearts of enthusiasts for generations to come.

Section 1.2: Conception of the 442

The birth of the Oldsmobile 442 was not an accident, but a deliberate response to a changing automotive landscape. In the early 1960s, a group of visionary engineers and designers at Oldsmobile recognized the growing demand for performance-oriented vehicles. This team, led by John Beltz, Dale Smith, and Bob Dorshimer, became the driving force behind what would soon become an icon in the muscle car world.

The initial design goals for the 442 were ambitious yet clear: create a high-performance vehicle that could compete with the likes of the Pontiac GTO while maintaining Oldsmobile's reputation for refinement and engineering excellence. The team aimed to strike a perfect balance between raw power and sophisticated handling, a combination that would set the 442 apart from its competitors.

The name "442" itself was a stroke of marketing genius, instantly communicating the car's key features to enthusiasts. Initially, it stood for a four-barrel carburetor, four-speed manual transmission, and dual exhausts. This simple yet effective naming convention would later evolve, but its original meaning encapsulated the car's performance-oriented nature.

Early prototypes of the 442 faced numerous challenges. The engineering team grappled with issues such as weight distribution, cooling system efficiency, and transmission durability. Each problem was methodically addressed, with solutions often pushing the boundaries of contemporary automotive technology. For instance, the team developed a unique cooling system to handle the increased heat generated by the high-performance engine, a feature that would become a hallmark of the 442's reliability.

The decision to green-light the 442 project was not taken lightly. Oldsmobile executives were initially hesitant, concerned about the potential impact on the brand's image and the associated financial risks. However, the persistence of the development team, coupled with growing market trends and the success of competitors like the Pontiac GTO, eventually convinced the higher-ups to take the plunge.

This decision-making process involved numerous meetings, presentations, and even a few late-night test drives, during which executives could experience the prototype's potential firsthand. Many Oldsmobile veterans still remember the moment when the project received final approval as a turning point for the brand.

The conception of the 442 was more than just the creation of a new car model; it represented Oldsmobile's commitment to innovation and its willingness to adapt to a changing market. The careful planning, engineering challenges, and corporate negotiations all contributed to the birth of a vehicle that would not only define Oldsmobile's performance heritage but also leave an indelible mark on the entire muscle car era.

As the 442 moved from concept to reality, the automotive world stood on the brink of a new era, one where Oldsmobile would prove it could build a muscle car to rival the best in the business. The stage was set for a legend to be born.

Section 1.3: The 1964 Debut

The year 1964 marked a pivotal moment in automotive history as Oldsmobile unveiled its answer to the growing demand for high-performance vehicles, the 442. Initially introduced as an option package for the Cutlass and F-85 models, the 442 represented Oldsmobile's bold entry into the burgeoning muscle car market.

The 442 designation was derived from its key features: a four-barrel carburetor, four-speed manual transmission, and dual exhaust. This combination promised enthusiasts a potent blend of power and performance that would soon become legendary. Under the hood, the 442 boasted a 330 cubic-inch V8 engine, specially tuned to produce an impressive 310 horsepower and 355 lb-ft of torque. This powerplant was a significant step up from the standard Cutlass offerings, immediately setting the 442 apart as a serious contender in the performance arena.

When the 442 hit showroom floors, it created quite a stir among automotive journalists and enthusiasts alike. Car and Driver magazine praised its "brilliant performance and handling," while Hot Rod called it "a factory hot rod with a warranty." The public reception was equally enthusiastic, with many buyers drawn to the 442's combination of Oldsmobile's reputation for quality and the newfound focus on high performance.

Sales figures for the first year were modest but promising. Oldsmobile sold approximately 2,999 442 packages, a respectable number for a mid-year introduction. This early success validated Oldsmobile's decision to enter the muscle car market and set the stage for increased production in the coming years.

Compared to its contemporaries, the 1964 442 held its own admirably. While the Pontiac GTO, often credited with kickstarting the muscle car era, outsold the 442, many enthusiasts preferred the Oldsmobile's more refined handling and overall balance. The 442 also compared favorably to other newcomers like the Chevrolet Chevelle SS and the Ford Fairlane Thunderbolt in terms of performance and build quality.

The 1964 debut of the 442 was more than just the introduction of a new car model; Oldsmobile declared that it was ready to compete in the high-stakes world of performance automobiles. The positive reception and promising sales figures of this first-year model laid the groundwork for what would become one of the most respected names in muscle car history. As the 1964 model year came to a close, it was clear that the 442 was not just a one-off experiment, but the beginning of a performance legacy that would define Oldsmobile for years to come.

Section 1.4: Rapid Evolution: 1965-1967

The years between 1965 and 1967 marked a period of rapid evolution for the Oldsmobile 442, solidifying its position as a formidable contender in the muscle car arena. Each model year brought significant improvements and changes, reflecting Oldsmobile's commitment to staying at the forefront of performance engineering.

In 1965, the 442 received its first significant update. The most notable change was the introduction of a larger, more powerful engine. The original 330 cubic inch V8 was replaced with a robust 400 cubic inch powerplant, significantly boosting the car's performance. This new engine, combined with a four-barrel carburetor, produced an impressive 345 horsepower and 440 lb-ft of torque. The '65 model also saw improvements in the suspension system, with a stiffer front stabilizer bar and revised shock absorbers enhancing the car's handling characteristics.

The 1966 model year brought further refinements to the 442. Oldsmobile introduced a new tri-carb setup as an option, pushing the horsepower rating to 360. This power increase, coupled with chassis improvements, made the '66 442 a serious threat on both the street and the drag strip. Aesthetically, the car underwent a minor facelift, featuring a more aggressive grille design and new taillights, which give it a fresher, more modern appearance.

The most significant change came in 1967 when the 442 was elevated from an option package to its own distinct model within the Oldsmobile lineup. This move signaled Oldsmobile's full commitment to the muscle car market and allowed for greater differentiation from the standard Cutlass. The '67 442 boasted a redesigned body with a longer, sleeker profile and a more luxurious interior, setting it apart from its predecessors and competitors alike.

Performance enhancements continued during this period, with each year seeing incremental improvements in power, handling, and braking. The introduction of the W-30 package in 1966, which continued into 1967, offered enthusiasts an even higher level of performance. This package included a hotter cam, improved carburetion, and functional ram-air induction, pushing the 442's performance capabilities to new heights.

As the 442 evolved, its reputation among enthusiasts and racers grew exponentially. Success on the drag strip and in various racing series helped cement its status as an actual performance machine. The car's combination of raw power, balanced handling, and Oldsmobile's reputation for reliability made it a favorite among those who demanded more than just straight-line speed.

By the end of 1967, the Oldsmobile 442 had transformed from a relatively unknown option package to one of the most respected names in the muscle car world. Its rapid evolution over these three years set the stage for even greater achievements to come, laying the foundation for what would become one of the most iconic muscle cars of all time.

Section 1.5: Engineering Marvels of the Early 442

The early Oldsmobile 442 was a testament to engineering excellence, pushing the boundaries of performance and design in the burgeoning muscle car era. At the heart of this mechanical marvel lay its powerplant, a robust V8 engine that set the standard for raw power and reliability. The initial 1964 model boasted a 330 cubic inch (5.4L) V8, producing an impressive 310 horsepower and 355 lb-ft of torque. This engine was not just about brute force; it incorporated advanced features for its time, such as a four-barrel carburetor and high-flow cylinder heads, allowing for efficient fuel delivery and combustion.

As the 442 evolved, so did its engine. By 1965, Oldsmobile had increased the displacement to 400 cubic inches (6.6L), further enhancing the car's performance credentials. This larger engine was capable of producing up to 345 horsepower, cementing the 442's reputation as a formidable contender in the muscle car arena. The engine's design incorporated innovative features such as a high-lift camshaft and free-flowing exhaust manifolds, maximizing power output while maintaining reliability.

Transmitting this power to the pavement was a choice of robust transmissions. Early 442s offered a standard three-speed manual transmission, but performance enthusiasts often opted for the available four-speed manual. This close-ratio gearbox allowed drivers to keep the engine in its power band, maximizing acceleration and driving excitement. For those who preferred a more relaxed driving experience, a two-speed Jetaway automatic transmission was also available, later replaced by a more efficient three-speed Turbo Hydra-Matic in 1967.

The driveline components of the 442 were engineered to handle the immense power and torque produced by its V8 engine. Heavy-duty drive shafts, robust differential gears, and reinforced axle shafts ensured that the 442 could repeatedly launch hard without fear of breakage. This attention to detail in the driveline engineering contributed significantly to the 442's reputation for durability under high-stress conditions.

One of the most notable engineering achievements of the early 442 was its advanced suspension system. Oldsmobile engineers recognized that straight-line speed was only part of the performance equation; handling and ride quality were equally important. The 442 featured a sophisticated suspension setup that included heavy-duty springs, specially tuned shock absorbers, and a front stabilizer bar. This configuration allowed for improved cornering capabilities without sacrificing ride comfort, a balance that many competitors struggled to achieve.

The braking system of the 442 was another area where engineering innovation shone. Recognizing the need for stopping power to match the car's acceleration, Oldsmobile equipped the 442 with larger, more effective brakes than those found in its standard models. The front drums were wider and featured finned aluminum hubs for improved heat dissipation. As the model progressed, options for even more capable braking systems became available, including front disc brakes that provided superior stopping power and fade resistance.

Aerodynamics and body design were also key considerations in the engineering of the early 442. While the primary focus of muscle cars was often straight-line performance, Oldsmobile engineers paid attention to the car's aerodynamic profile. The sleek body lines of the 442 were not just for aesthetics; they helped reduce drag and improve high-speed stability. Features like the distinctive hood scoops were functional as well as stylish, helping to direct cool air to the engine compartment.

The attention to detail in the 442's engineering extended to its interior as well. The cockpit was designed with the driver in mind, featuring easy-to-read gauges, comfortable seating, and well-placed controls. This driver-centric approach ensured that the 442 was not only powerful but also comfortable and enjoyable to drive for extended periods of time.

In conclusion, the engineering marvels of the early Oldsmobile 442 set it apart from its contemporaries. From its potent V8 engine and robust drivetrain to its advanced suspension and braking systems, every aspect of the 442 was carefully designed and executed. This holistic approach to performance engineering created a muscle car that was more than the sum of its parts, a true icon of American automotive ingenuity.

Section 1.6: Marketing and Promotion

Oldsmobile's marketing strategy for the 442 was a masterclass in automotive promotion, perfectly capturing the spirit of the muscle car era. From the outset, the company recognized the need to position the 442 as more than just another performance option; it had to be a lifestyle choice, a symbol of power and freedom for the discerning enthusiast.

The marketing team at Oldsmobile crafted a series of memorable ad campaigns that would resonate with their target audience. One of the most iconic slogans was "This One's Got Hustle!" which encapsulated the 442's performance prowess and street credibility. Another standout tagline was "Keeper of the Cool," emphasizing the car's blend of performance and style. These catchy phrases appeared in print ads, billboards, and television commercials, embedding themselves in the public consciousness.

Dealerships played a crucial role in promoting the 442. Oldsmobile provided extensive training to sales staff, ensuring they could effectively communicate the car's unique features and performance capabilities. Special showroom displays were created, highlighting the 442's muscular stance and premium components. Some dealerships even organized local drag racing events, allowing potential customers to experience the 442's power firsthand.

To create buzz and exclusivity, Oldsmobile introduced several special editions and promotional packages. The "W-30" option, for instance, became legendary among performance enthusiasts. These limited-run models often featured unique color schemes, enhanced performance parts, and exclusive badging, driving demand and collectibility.

Oldsmobile's marketing efforts specifically targeted young performance enthusiasts. The company sponsored youth-oriented events, partnered with popular musicians, and advertised in magazines catering to the car culture. They understood that winning over this demographic was key to building a loyal customer base and securing the 442's future success.

Building brand loyalty was a core focus of Oldsmobile's strategy. They fostered a sense of community among 442 owners through owners' clubs, newsletters, and exclusive events. These initiatives not only strengthened the bond between customers and the brand but also created a network of enthusiasts who became unofficial ambassadors for the 442.

The marketing team also leveraged the 442's racing success in their promotions. Victories on the track were prominently featured in advertisements, reinforcing the car's performance credentials. They highlighted the direct link between race-winning technology and the production models available to the public, appealing to those who wanted a piece of that racing heritage.

Oldsmobile's marketing efforts extended beyond traditional channels. They collaborated with model car manufacturers to produce scale replicas of the 442, ensuring that even those too young to drive could experience the allure of the brand. This strategy helped plant the seeds of desire in future customers.

Throughout its early years, the marketing and promotion of the Oldsmobile 442 set new standards in automotive advertising. By combining compelling messaging, targeted campaigns, and community-building initiatives, Oldsmobile not only sold cars but created a lasting legacy. The 442 became more than just a high-performance option; it became a cultural icon, a status symbol, and a dream car for a generation of enthusiasts. This marketing success played a crucial role in cementing the 442's place in muscle car history and continues to influence how performance cars are promoted to this day.

Section 1.7: Early Racing Success

The Oldsmobile 442's journey from showroom to racetrack was a natural progression that solidified its reputation as a true muscle car contender. As soon as the 442 hit the streets, enthusiasts and professional racers alike recognized its potential for domination on the track.

The 442's first forays into organized racing were nothing short of impressive. In 1965, just a year after its debut, the 442 made its mark in the NHRA's A/Stock class. The car's robust engine and finely-tuned suspension allowed it to compete with, and often outperform, more established muscle car names.

One of the most notable races that put the 442 on the map was the 1966 Trans-Am series. While not purpose-built for this type of racing, the 442 held its own against purpose-built pony cars, showcasing its versatility and raw power. Its performance in the series caught the attention of both fans and competitors, proving that Oldsmobile had created something truly special.

The 442's success wasn't limited to sanctioned races. It quickly became a favorite among drag racers, dominating local strips across the country. Its combination of torque and traction made it a formidable opponent in quarter-mile sprints, often leaving more famous muscle cars in its wake.

Several famous drivers became associated with the early 442, further cementing its reputation. One such driver was "Doc" Dick Oldham, who campaigned a 442 in A/Stock drag racing with great success. His victories helped spread the word about the 442's capabilities and drew more attention to the Oldsmobile brand in racing circles. Another notable name was Bob Senneker, who raced a 442 in the ARCA series. Senneker's success with the 442 demonstrated the car's potential in stock car racing, a discipline typically dominated by other brands.

The 442's racing success had a profound influence on its further development. Engineers at Oldsmobile paid close attention to feedback from racers, using their insights to refine and improve the car. This racing-derived knowledge led to enhancements in areas such as engine cooling, brake performance, and suspension tuning, all of which found their way into production models.

These improvements created a virtuous cycle: better performance on the track led to improved street cars, which in turn created more interest from racers. This continuous loop of development and refinement played a crucial role in building the 442's performance reputation.

As word spread of the 442's prowess on the track, its status among enthusiasts grew exponentially. It was no longer seen as just another muscle car option, but as a serious performance machine capable of going toe-to-toe with the best Detroit had to offer. The early racing success of the 442 laid the groundwork for its legendary status, proving that Oldsmobile could build cars that were not just fast in a straight line, but actual all-around performers.

The track became a proving ground where the 442 demonstrated its mettle, pushing the boundaries of what was possible with a production-based car. This racing heritage would continue to influence the 442's development throughout its lifespan, ensuring that it remained at the forefront of performance car engineering.

In the end, the early racing success of the Oldsmobile 442 did more than just win trophies. It established a legacy of performance that would define the model for years to come, inspiring a generation of enthusiasts and cementing its place in the pantheon of great American muscle cars.

High Octane Heritage: *Celebrating the Oldsmobile 442*

Chapter 2: Engineering Marvels: The Heart of the 442

Section 2.1: The Birth of the 442 Engine

The story of the Oldsmobile 442 engine begins with a simple yet powerful concept: create a high-performance powertrain that would set a new standard in the burgeoning muscle car market. The "442" designation, which would become synonymous with raw power and engineering excellence, originally stood for four-barrel carburetor, four-speed manual transmission, and dual exhausts. This combination was first offered as an option package for the F-85 and Cutlass models in 1964, marking the birth of a legend.

The initial 442 engine was a marvel of its time. Based on the Oldsmobile 330 cubic inch V8, it was bored out to 330 cubic inches and fitted with a higher-lift camshaft, stronger valve springs, and a modified combustion chamber design. These enhancements allowed the engine to produce an impressive 310 horsepower and 355 lb-ft of torque, figures that would make any car enthusiast's heart race.

Key engineers, including John Beltz, Dale Smith, and Bob Dorshimer, were instrumental in bringing the 442 engine to life. Their vision was to create a powerplant that could compete with the best from Pontiac and Chevrolet while maintaining Oldsmobile's reputation for reliability and sophistication. They worked tirelessly to balance performance with drivability, ensuring that the 442 engine would be as comfortable on the street as it was formidable on the drag strip.

Compared to contemporary Oldsmobile engines, the 442 was in a class of its own. While the standard Oldsmobile V8s were known for their smooth operation and decent power output, the 442 engine pushed the boundaries of what was possible. It featured a higher compression ratio, more aggressive camshaft profiles, and larger valves, all of which contributed to its significant power advantage over its siblings.

The early reception of the 442 engine was nothing short of enthusiastic. Automotive journalists praised its smooth power delivery and impressive acceleration. In road tests, cars equipped with the 442 engine consistently outperformed their competitors, achieving 0-60 mph times in the mid-6-second range, a blistering performance for the mid-1960s.

Performance metrics told only part of the story, however. What truly set the 442 engine apart was its versatility. It was equally at home powering a family sedan for a Sunday drive as it was propelling a drag racer down the quarter-mile. This dual nature, refined yet powerful, would become a hallmark of the 442 engine throughout its lifespan.

The birth of the 442 engine marked a turning point for Oldsmobile. It signaled the division's commitment to performance and innovation, setting the stage for years of engineering advancements that would keep the 442 at the forefront of the muscle car era. As we'll see in the following sections, this was just the beginning of a legendary journey in automotive engineering.

Section 2.2: Evolution of Power

The Oldsmobile 442 engine underwent a remarkable transformation throughout its lifespan, steadily evolving into one of the most formidable powerplants of the muscle car era. This journey of continuous improvement began with the original 1964 model and culminated in the peak years of the late 1960s and early 1970s.

Initially, the 442 package was built around a 330 cubic inch V8 engine, producing a respectable 310 horsepower. However, Oldsmobile engineers quickly realized the potential for greater performance. In 1965, they increased the engine displacement to 400 cubic inches, boosting output to 345 horsepower. This significant upgrade set the stage for the 442's reputation as a serious contender in the muscle car arena.

The introduction of the W-30 package in 1966 marked a pivotal moment in the 442's evolution. This high-performance option included a more aggressive camshaft, improved carburetion, and a low-restriction air intake system. These enhancements pushed the engine's output to 360 horsepower, cementing the 442's status as a top-tier muscle car.

As the years progressed, Oldsmobile continued to refine and enhance the 442 engine. In 1968, the adoption of improved cylinder heads and a redesigned intake manifold led to a significant increase in horsepower to 365. The following year saw another considerable leap with the introduction of the 455 cubic inch V8, which produced a massive 380 horsepower and an astounding 500 lb-ft of torque.

The pinnacle of 442 performance came in 1970 with the W-30 455 engine. This powerhouse produced an advertised 370 horsepower, though many experts believe the actual output was closer to 400 horsepower. The engine's enormous torque output, peaking at 500 lb-ft, made the 442 a force to be reckoned with on both the street and the drag strip.

However, the early 1970s brought new challenges in the form of stricter emission standards and a shift in market preferences towards more fuel-efficient vehicles. Oldsmobile engineers rose to the occasion, developing innovative solutions to maintain performance while meeting new regulatory requirements. They implemented technologies such as exhaust gas recirculation (EGR) systems and catalytic converters, carefully tuning these components to minimize their impact on engine output.

Despite these challenges, the 442 engine continued to evolve. The introduction of electronic ignition systems improved reliability and performance, while advancements in metallurgy enabled the development of lighter, yet stronger, engine components. Even as horsepower ratings declined on paper due to changes in measurement standards and emissions equipment, Oldsmobile worked tirelessly to maintain the 442's reputation for strong, usable power.

Throughout its lifespan, the 442 engine's evolution was characterized by a relentless pursuit of performance balanced with adaptability to changing regulations and market demands. From its humble beginnings as a high-output option for the F-85, to its peak as one of the most powerful engines of the muscle car era, to its later years as a refined and efficient powerplant, the 442 engine remained at the forefront of automotive engineering.

This continuous evolution not only kept the 442 competitive in a rapidly changing market but also ensured its place in automotive history as one of the most revered engines of its time. The legacy of the 442 engine's development continues to inspire enthusiasts and engineers alike, serving as a testament to the ingenuity and passion of Oldsmobile's engineering team.

Section 2.3: Innovative Engineering Techniques

The Oldsmobile 442 engine was a testament to innovative engineering, incorporating several cutting-edge techniques that set it apart from its competitors. At the heart of its design was a unique cylinder head configuration that maximized airflow and combustion efficiency. The engineers at Oldsmobile developed a high-flow port design that allowed for improved fuel-air mixture delivery to the combustion chambers. This, coupled with larger valves and optimized combustion chamber shapes, resulted in superior power output and throttle response.

The valve train technology in the 442 engine was equally advanced for its time. Oldsmobile implemented a robust rocker arm system with higher-than-average ratios, enabling greater valve lift without compromising reliability. This design choice enabled the engine to breathe more efficiently at high RPMs, a crucial factor in its impressive performance. Additionally, the use of hardened valve seats and stems increased durability, allowing the engine to withstand the rigors of high-performance driving.

Fuel delivery was another area where the 442 engine showcased innovative engineering. As carburetors evolved, Oldsmobile kept pace by implementing increasingly sophisticated systems. The introduction of the Rochester Quadrajet carburetor was a game-changer, offering the benefits of a small primary bore for improved fuel economy and drivability, combined with large secondary bores for maximum power output. This dual-stage design provided the best of both worlds, making the 442 engine both tractable for daily driving and ferocious when pushed to its limits.

Cooling system innovations played a crucial role in maintaining the 442 engine's performance and longevity. Oldsmobile engineers developed a high-capacity cooling system that included a larger radiator, an improved water pump, and optimized coolant flow paths through the engine block and heads. This advanced cooling architecture allowed the engine to maintain optimal operating temperatures even under extreme conditions, preventing power loss due to heat soak and reducing the risk of engine damage.

Noise, vibration, and harshness (NVH) reduction was an often-overlooked aspect of muscle car engineering, but Oldsmobile paid special attention to this area in the 442 engine. The use of hydraulic lifters not only reduced maintenance requirements but also significantly quieted valve train noise. Furthermore, the engine mounts were carefully designed to isolate engine vibrations from the chassis, providing a smoother driving experience without compromising performance.

One of the most innovative features of the 442 engine was its forced air induction system, particularly in the W-30 package. This system utilized a fiberglass hood with functional air scoops that fed cold air directly into the carburetor. The implementation of this "Outside Air Induction" system was a stroke of engineering genius, boosting power output by providing denser, cooler air to the engine while also creating a distinctive and aggressive appearance.

Oldsmobile's engineers also pioneered the use of forged components in a mass-produced engine. The 442's crankshaft, connecting rods, and pistons were forged rather than cast, providing exceptional strength and durability. This allowed the engine to withstand higher cylinder pressures and RPMs, contributing to its reputation for reliability even under extreme use.

The cumulative effect of these innovative engineering techniques was an engine that not only produced impressive power figures but also offered a level of refinement and durability that was rare in the muscle car era. The 442 engine's ability to deliver exhilarating performance while maintaining reasonable fuel efficiency and reliability for daily use was a testament to the forward-thinking approach of Oldsmobile's engineering team. Their work on the 442 engine pushed the boundaries of what was possible in a production car engine and left a lasting impact on automotive engineering.

Section 2.4: Transmission and Drivetrain

The heart of the Oldsmobile 442 may have been its powerful engine, but the transmission and drivetrain played crucial roles in translating that raw power into exhilarating performance on the road. Throughout the 442's production run, Oldsmobile engineers continually refined these components to enhance the car's overall driving experience.

Manual transmission options were a key focus for performance enthusiasts. Early 442 models offered a robust three-speed manual as standard, but it was the introduction of the four-speed manual that truly excited drivers. This transmission, sourced from Muncie, offered crisp shifts and improved gear ratios for enhanced acceleration and top speed. As the 442 evolved, so did its manual transmission options. Later models featured enhanced synchronizers and stronger gears to accommodate the increasing power output of the engine.

While purists favored manual transmissions, the automatic transmission market was not neglected. Oldsmobile's engineers worked tirelessly to improve the performance of their automatic offerings. The Turbo Hydra-Matic 400, introduced in the mid-1960s, was a game-changer. This three-speed automatic transmission was renowned for its durability and smooth operation. Engineers fine-tuned its shift points and torque converter characteristics to maximize the 442's performance potential. In later years, electronic controls were integrated to enhance shift quality and efficiency further.

The rear axle and differential were also subjects of continuous improvement. Recognizing the importance of putting power to the pavement, Oldsmobile offered a range of rear axle ratios to suit different driving preferences. Performance-oriented drivers could opt for lower numerical ratios for quicker acceleration, while those seeking better fuel economy and highway cruising could choose higher ratios. The introduction of limited-slip differentials was another significant upgrade, providing better traction and handling, especially in high-performance driving scenarios.

Driveshaft and U-joint enhancements were crucial in handling the 442's increasing power output. Engineers employed stronger materials and improved designs to reduce vibration and increase durability. Some models featured balanced driveshafts to minimize NVH (Noise, Vibration, and Harshness) at high speeds, contributing to a smoother driving experience.

Power delivery optimization was an ongoing process throughout the 442's lifespan. Engineers worked to minimize drivetrain losses, ensuring that as much of the engine's power as possible made it to the rear wheels. This involved everything from refining gear tooth profiles to improving lubrication systems for better efficiency under high-stress conditions.

One notable innovation was the introduction of the Hurst/Olds package in 1968. This collaboration between Oldsmobile and Hurst Performance resulted in a specially tuned drivetrain that included a unique shifter for the manual transmission, providing shorter throws and more precise gear changes. This package became legendary among enthusiasts for its performance-enhancing capabilities.

As emission regulations tightened in the 1970s, Oldsmobile engineers faced new challenges in maintaining the 442's performance while meeting legal requirements. This led to further refinements in the drivetrain, including the introduction of more efficient torque converters and revised gear ratios to balance power and economy.

The transmission and drivetrain developments in the Oldsmobile 442 were not just about raw performance; they also focused on durability and driver comfort. The engineering team strived to create a powertrain that could handle the stresses of high-performance driving while still providing a smooth, enjoyable experience for daily use. This balance of performance and practicality was a key factor in the 442's enduring appeal.

In retrospect, the evolution of the 442's transmission and drivetrain is a testament to Oldsmobile's commitment to continuous improvement. From manual to automatic, from the driveshaft to the differential, each component was scrutinized and enhanced to create a cohesive, high-performance driving experience. These advancements not only defined the 442 but also influenced drivetrain development across the muscle car era and beyond.

Section 2.5: Pushing the Limits: Racing Modifications

The Oldsmobile 442's reputation as a muscle car icon was cemented not only on the streets but also on the racetrack. This section delves into the various racing modifications that pushed the 442 engine to its absolute limits, showcasing its true potential and influencing future developments.

Factory racing upgrades played a crucial role in enhancing the 442's performance. Oldsmobile engineers developed a range of high-performance parts specifically for racing applications. These included forged pistons, stronger connecting rods, and high-flow cylinder heads. The W-30 package, introduced in 1966, was a prime example of factory racing upgrades. It featured a fiberglass hood with functional air scoops, a more aggressive camshaft, and a high-flow air induction system. These modifications significantly boosted the engine's output and responsiveness, giving racers a competitive edge right off the showroom floor.

The aftermarket performance industry also embraced the 442, offering a plethora of enhancements for enthusiasts and racers alike. Companies like Edelbrock and Holley developed specialized intake manifolds and carburetors that could extract even more power from the already potent engine. High-performance camshafts, headers, and exhaust systems became popular upgrades, allowing owners to tailor their 442's performance to their specific needs. Some of the more extreme modifications included nitrous oxide systems and supercharger kits, which could dramatically increase horsepower for drag racing applications.

Notable racing engine builds took the 442 powerplant to extraordinary levels. One famous example was the engine developed for the 1970 Oldsmobile F-85 that competed in the NHRA Super Stock class. This highly modified 455 cubic inch V8 produced over 600 horsepower, a staggering figure for the time. Another legendary build was the 442 engine used in the Trans-Am series, which featured advanced porting and flowing techniques, resulting in exceptionally high-rpm performance.

Dyno results and real-world performance of these modified 442 engines were often astonishing. Some of the most potent builds were capable of producing over 700 horsepower on the dynamometer, translating to quarter-mile times in the low 10-second range. On road courses, modified 442 engines demonstrated impressive durability and consistent power delivery, allowing drivers to compete effectively against purpose-built race cars.

The influence of these racing modifications on future Oldsmobile engines cannot be overstated. Many of the lessons learned from pushing the 442 engine to its limits were incorporated into subsequent production models. The development of more robust internal components, improved oiling systems, and advanced cylinder head designs all stemmed from racing experience. Even as emissions regulations began to tighten in the early 1970s, the knowledge gained from racing helped Oldsmobile engineers maintain performance while meeting new standards.

The racing modifications applied to the 442 engine not only enhanced its performance but also contributed to its legendary status. They demonstrated the inherent strength and potential of the basic design, proving that with the right upgrades, the 442 could compete with the best performance cars of its era. This legacy of performance and adaptability continues to inspire enthusiasts and engineers alike, ensuring that the 442 engine remains a revered piece of muscle car history.

Section 2.6: Engineering Challenges and Solutions

The journey of the 442 engine was not without its share of obstacles and hurdles. As with any groundbreaking engineering feat, the Oldsmobile team faced numerous challenges in their quest to create and maintain the perfect muscle car powerplant. One of the most significant challenges was striking the delicate balance between performance and reliability. The engineers had to push the boundaries of power output while ensuring that the engine could withstand the rigors of daily driving and occasional high-performance use.

To achieve this balance, the team employed innovative design techniques and materials. They focused on strengthening critical components such as the crankshaft, connecting rods, and pistons to handle the increased stresses of high-performance operation. Advanced metallurgy played a crucial role, with the introduction of nodular iron and forged steel components that offered superior strength without excessive weight gain.

As the 442 engine evolved, specific weaknesses and common issues emerged. Early models sometimes suffered from oil consumption problems and valve train wear. The engineering team responded swiftly, implementing improved piston ring designs and more durable valve guides. They also developed enhanced oiling systems to ensure proper lubrication under high-stress conditions, a critical factor in maintaining engine longevity.

The rapid pace of technological advancement in the automotive industry presented another ongoing challenge. The Oldsmobile engineers had to innovate to stay ahead of the curve continually. This led to the adoption of new manufacturing techniques, such as precision computer-controlled machining, which allowed for tighter tolerances and improved engine efficiency.

Changing market demands also posed significant challenges. As fuel prices rose and environmental concerns gained prominence, the team had to adapt the 442 engine to meet new expectations without sacrificing its performance heritage. This led to the development of more efficient fuel delivery systems and the gradual integration of emissions control technologies that maintained power while reducing harmful exhaust outputs.

One of the most formidable challenges was the intense competition from other muscle cars of the era. Rivals like the Pontiac GTO, Chevrolet Chevelle SS, and Ford Mustang were constantly upping the ante in the horsepower wars. The Oldsmobile team responded by continually refining the 442 engine, introducing performance packages like the W-30, and pushing the boundaries of what was possible within GM's corporate guidelines.

To overcome these challenges, the engineering team fostered a culture of innovation and problem-solving. They maintained close relationships with racers and performance enthusiasts, using real-world feedback to inform their development process. This approach allowed them to quickly identify and address issues, often before they became widespread problems.

The team also leveraged Oldsmobile's racing program as a testbed for new technologies and performance enhancements. Innovations proven on the track often found their way into production models, ensuring that the 442 engine remained at the forefront of muscle car performance.

In the face of tightening emissions regulations, the engineers got creative. They developed more efficient combustion chamber designs and experimented with early forms of electronic engine management to maximize power while minimizing pollutants. This forward-thinking approach helped the 442 engine remain relevant and robust even as many of its competitors began to lose their edge.

Ultimately, the story of the 442 engine's development is one of persistence, ingenuity, and adaptation. The Oldsmobile engineering team's ability to overcome a myriad of challenges not only resulted in a legendary powerplant but also contributed significantly to the advancement of automotive engineering as a whole. Their solutions to these complex problems set new standards in the industry and helped cement the 442's place in muscle car history.

Section 2.7: Legacy of the 442 Engine

The Oldsmobile 442 engine left an indelible mark on the automotive world, its influence extending far beyond its years of production. This powerplant's legacy is evident in the engines that followed, not just within Oldsmobile but across the entire General Motors lineup. The innovative engineering solutions developed for the 442 engine served as a blueprint for future high-performance engines, setting new standards for power, efficiency, and reliability.

The impact of the 442 engine on future Oldsmobile and GM engines cannot be overstated. Many of the advancements made during its development, such as improved cylinder head designs and advanced valve train technology, were incorporated into subsequent engine families. The lessons learned from pushing the limits of performance while maintaining streetability were invaluable, informing the design of engines well into the 21st century.

Beyond its direct influence on GM products, the 442 engine played a significant role in shaping the overall development of muscle car engineering. It raised the bar for what was possible in a production vehicle, forcing competitors to innovate and improve their own offerings. The 442 engine's ability to deliver impressive power while meeting increasingly stringent emissions standards set a benchmark that other manufacturers strived to match.

Today, the 442 engine holds a special place in the hearts of collectors and enthusiasts. Its rarity and historical significance have made it highly sought after, with pristine examples commanding premium prices. However, this desirability comes with unique restoration challenges. The scarcity of original parts and the specialized knowledge required to rebuild these engines properly have created a niche market for experts in 442 engine restoration.

Interestingly, the legacy of the 442 engine lives on not just in preservation but in evolution. Modern performance upgrades have allowed enthusiasts to breathe new life into these classic powerplants. Advanced fuel injection systems, stronger internal components, and computer-controlled engine management have enabled 442 engines to produce power figures that would have been unimaginable in the 1960s, all while maintaining reliability and drivability.

In the grand tapestry of automotive history, the 442 engine occupies a prominent position. It represents a pivotal moment in the muscle car era, embodying the spirit of innovation and the pursuit of performance that defined that period. More than just a relic of the past, the 442 engine continues to inspire engineers and enthusiasts alike, serving as a reminder of what can be achieved when passion and engineering excellence combine.

The 442 engine's legacy is not just about raw power or impressive statistics. It's about the human ingenuity that went into its creation, the joy it brought to drivers, and the way it pushed the entire automotive industry forward. As we look back on the 442 engine, we see more than just a mechanical marvel; we see a testament to the enduring appeal of American muscle and the timeless allure of a well-engineered machine.

High Octane Heritage: *Celebrating the Oldsmobile 442*

Chapter 3: Design Evolution: Styling Changes Through the Decades

Section 3.1: The Birth of an Icon (1964-1967)

The Oldsmobile 442's journey began in 1964, not as a standalone model, but as an innovative option package for the Cutlass. This subtle start would soon evolve into one of the most iconic muscle cars of the era, setting the stage for a design legacy that would span decades.

When the 442 package first appeared, it was distinguished primarily by subtle badging and dual exhaust. The understated approach was intentional, as Oldsmobile was testing the waters of the burgeoning muscle car market. The 1964 model's design was clean and elegant, with smooth lines and a relatively conservative appearance that belied its performance capabilities. This initial offering was a testament to Oldsmobile's ability to blend performance with sophistication, a trait that would become a hallmark of the 442 throughout its lifespan.

As the muscle car era began to take shape, the 1965 model year saw the 442 embrace a more aggressive aesthetic. The introduction of the iconic bulged hood signaled the 442's muscle car intentions, giving the car a more imposing presence on the road. This design element not only enhanced the car's visual appeal but also served a functional purpose, accommodating the larger engine and hinting at the power lurking beneath. The front grille became more pronounced, and the overall stance of the car appeared more athletic, setting it apart from its more pedestrian Cutlass siblings.

The years 1966 and 1967 saw Oldsmobile refining the 442 formula, both in terms of performance and design. The 1967 model year further sharpened the 442's aggressive look with the introduction of distinctive louvered hood inserts. This functional and stylish element helped set it apart and signaled the serious performance hardware that lay beneath. These inserts not only added a touch of aggression to the car's appearance but also improved engine cooling, marrying form and function in a way that typified muscle car design of the era. The body lines were further accentuated, with a more pronounced Coke-bottle shape that emphasized the car's power and speed even when stationary.

Color and trim options played a crucial role in defining the 442's identity during these early years. Vibrant colors like Trumpet Gold and Ember Red became synonymous with the early 442s, allowing owners to make a bold statement on the street or at the drag strip. These eye-catching hues were often paired with contrasting stripes or blacked-out elements, further enhancing the car's sporting character. Interior options were equally important, with bucket seats, sports steering wheels, and performance-oriented instrumentation becoming staples of the 442 package.

The design evolution of the 442 during this period was not just about aesthetics; it was a reflection of the car's increasing performance capabilities and Oldsmobile's commitment to the muscle car segment. A lead designer at Oldsmobile during this era, recalls,

"We wanted the 442 to look fast even when standing still. Every curve, every line was designed to convey power and speed."

This philosophy is evident in the progression of the 442's design from 1964 to 1967. What started as a subtle performance package grew into a visually distinct model that commanded attention. The evolution of the 442's design during these formative years laid the groundwork for what would become one of the most respected names in the muscle car pantheon.

As the 1960s progressed, the 442's design would continue to evolve, becoming bolder and more distinctive. But it was these early years that established the 442's design language – a perfect blend of Oldsmobile sophistication and raw muscle car aggression. This unique identity would carry the 442 through the heights of the muscle car era and beyond, cementing its place in automotive design history.

Section 3.2: The Golden Age (1968-1972)

The years 1968 to 1972 marked the pinnacle of the Oldsmobile 442's design evolution, a period often referred to as the Golden Age of muscle cars. This era saw the 442 transform from a respected performance package into an automotive icon, with styling that perfectly captured the spirit of the times.

The 1968 model year ushered in a revolutionary redesign that would define the 442's most iconic look. Gone were the relatively conservative lines of the earlier models, replaced by a more aggressive, sculpted body style that exuded power and speed. The new design featured a long hood, short deck proportions that emphasized the car's performance potential. The front end was dominated by a bold, split grille flanked by quad headlights, giving the 442 a menacing face that commanded attention on the road.

As the 442 entered the 1970s, its design reached its zenith. The 1970 model, in particular, is often considered the pinnacle of 442 styling. It featured muscular fender flares that accentuated the car's width and stance, conveying a sense of raw power. The distinctive hood bulge, now larger and more pronounced, hinted at the potent engine lurking beneath. Chrome trim was used judiciously, adding just the right amount of flash without overwhelming the car's purposeful appearance.

During this period, Oldsmobile designers introduced several distinctive features that set the 442 apart from its competitors. The W-30 package, for instance, added eye-catching red inner fender wells, visible through the functional hood scoops. This not only served a practical purpose by directing cool air to the engine but also became a signature visual element of high-performance 442s.

The interior of the 442 also saw significant innovations during this golden age. In 1969, Oldsmobile introduced the famous "Tic-Toc-Tach," an ingenious combination of analog clock and tachometer in a single gauge. This unique instrument perfectly blended functionality with style, becoming an instant classic and a much-sought-after feature among enthusiasts.

Color and trim options during this period were bold and expressive, reflecting the exuberant spirit of the late '60s and early '70s. Vibrant hues like Twilight Blue, Ram Rod Red, and Sebring Yellow were popular choices, often paired with contrasting stripes or vinyl tops. Interior color options were equally daring, with bright reds, blues, and even greens available alongside more traditional black and white.

The Golden Age also saw the introduction of several limited editions and special models that pushed the boundaries of 442 design even further. The most notable of these was the Hurst/Olds, a collaboration between Oldsmobile and Hurst Performance. The 1969 Hurst/Olds, with its distinctive gold and white paint scheme, twin-scoop hood, and trunk-mounted wing, represented the ultimate expression of muscle car excess.

As muscle car mania reached its peak in the early 1970s, the 442's design continued to evolve. The 1971 and 1972 models featured a more refined look, with a redesigned front end that incorporated a larger grille and a more substantial bumper assembly. These changes gave the car a slightly more mature appearance without sacrificing its performance image.

Throughout this golden era, Oldsmobile designers struck a perfect balance between form and function. Every curve, every line of the 442 served a purpose, whether it was to improve aerodynamics, enhance cooling, or simply to turn heads. The result was a car that looked fast even when standing still, a true embodiment of the muscle car ethos.

The design language developed during this period would influence Oldsmobile styling for years to come, cementing the 442's place in automotive design history. Even as changing regulations and market preferences would soon force dramatic changes in the automotive landscape, the 442 designs from 1968 to 1972 remain timeless classics, beloved by enthusiasts and collectors to this day.

Section 3.3: Adapting to Change (1973-1977)

As the 1970s progressed, the automotive landscape underwent significant shifts, and the Oldsmobile 442 was not immune to these changes. The era of unbridled muscle car performance was coming to an end, and automakers had to adapt to new regulations and changing consumer preferences. This period saw the 442 evolve from a raw performance machine to a more refined grand tourer, all while maintaining its distinctive character.

The impact of regulations on the 442's design was most evident in its front-end styling. The introduction of 5-mph bumpers in 1973 significantly altered the car's appearance. These larger, more prominent bumpers were designed to withstand low-speed collisions without damage, but they posed a challenge for designers attempting to maintain the 442's sleek profile. Oldsmobile's solution was to integrate these bumpers as seamlessly as possible, using chrome accents and careful contouring to blend them into the overall design. While this change was noticeable, it was handled more gracefully than on some competing models of the era.

The years 1973 and 1974 marked the end of an era for the 442, as these were the last models built on the full-size A-body platform. The 1973 model featured a more formal roofline, reflecting a shift in consumer preferences towards a more luxurious appearance. This shift was evident in the car's overall proportions, with a longer hood and a more upright greenhouse. The iconic dual headlights were retained, but they were now set in a more prominent grille that extended the full width of the car, giving it a broader, more imposing presence on the road.

In 1975, the 442 transitioned to the new Colonnade body style, marking a significant departure from its previous designs. The Colonnade era, which lasted until 1977, introduced a distinctive "cockpit" style interior that wrapped around the driver, creating a more intimate driving environment. Externally, the car featured a sleeker profile with a faster C-pillar and larger glass areas, improving visibility while maintaining a sporty appearance.

This period saw Oldsmobile designers balancing performance aesthetics with luxury touches, reflecting the 442's evolution from a pure muscle car to a more sophisticated performance coupe. Plush interiors became a hallmark of this era, with high-quality materials and more comfortable seating. The exterior lines became softer and more flowing, a departure from the aggressive, chiseled look of earlier models. This change was not just aesthetic but also functional, as the smoother lines contributed to improved aerodynamics and fuel efficiency.

The color and trim options available for the 442 also evolved during this time. While the bold, eye-catching colors of the muscle car era were still prevalent, a noticeable shift towards more subdued earth tones and metallic finishes emerged. Colors like Russet Metallic, Glacier Blue, and Burgundy became popular choices, reflecting the more sophisticated image the 442 was cultivating. Interior color schemes followed suit, with rich, deep tones replacing the brighter hues of the previous decade.

Despite these changes, Oldsmobile worked to maintain key visual cues that identified the 442 as a performance model. Special graphics packages, hood bulges, and unique wheel designs helped differentiate the 442 from standard Cutlass models. The iconic "442" badging remained, though it was often incorporated in more subtle ways than in previous years.

The 1973-1977 period was one of significant transition for the Oldsmobile 442. It navigated the challenges of stricter regulations and shifting market demands while maintaining its performance heritage. The design changes during this era reflected a broader trend in the American automotive industry – the move towards more comfortable, efficient, and sophisticated vehicles. While some purists may have lamented the departure from the raw muscle car aesthetics of the 1960s, these changes allowed the 442 to remain relevant and desirable in a rapidly changing market.

This era of the 442 demonstrated Oldsmobile's ability to adapt its flagship performance model to new realities while retaining much of what made it special. The 442 of the mid-1970s may have been a different animal from its predecessors, but it continued to embody the spirit of American performance, albeit in a more refined package.

Section 3.4: The Downsized Years (1978-1980)

The late 1970s marked a significant turning point in the automotive industry, and the Oldsmobile 442 was not immune to the sweeping changes. As fuel efficiency concerns and stricter emissions regulations took center stage, the once-mighty muscle car found itself adapting to a new era. The result was a downsized 442 that, while smaller in stature, still aimed to maintain its performance pedigree.

In 1978, the 442 underwent a dramatic transformation, transitioning to the more petite Cutlass body. This new compact 442 adopted a crisp, angular design language that was a stark departure from its muscular predecessors. The sleek, wedge-shaped profile was a nod to the changing automotive aesthetics of the time, emphasizing efficiency and aerodynamics over brute force.

Despite the size reduction, Oldsmobile's designers worked diligently to ensure that the 442 retained its unique identity. Distinctive stripes and badging played a crucial role in maintaining the 442's performance image. The iconic "442" emblem, now more subtle than in previous years, still adorned the fenders and deck lid, serving as a reminder of the car's storied past.

The front end of the downsized 442 featured a more upright grille flanked by rectangular headlights, a common design element of the era. As the model years progressed, subtle changes were made to enhance both aesthetics and functionality. The 1980 model, for instance, introduced a sloping front end that improved aerodynamics, a clear nod to the growing importance of fuel efficiency.

Inside the cabin, the 442 continued to evolve. The interior design reflected a blend of sportiness and comfort, with high-back bucket seats and a driver-oriented dashboard. One of the most notable interior innovations came in 1979 with the introduction of an optional digital dashboard. This electronic display brought the 442 into the modern age, offering a futuristic touch that appealed to tech-savvy buyers.

Color and trim options during this period reflected the changing tastes of the American consumer. While bold colors were still available, there was a noticeable shift towards more subdued metallic tones and earth hues. Interior color schemes followed suit, offering a mix of classic dark tones and more contemporary options.

The final years of the rear-wheel-drive 442 saw a return to a more aggressive front fascia design. The 1980 model featured a bolder grille and headlight treatment, almost as if the 442 was making one last visual statement before the end of an era. This design served as a fitting send-off for the traditional rear-wheel-drive platform that had defined the 442 for over a decade.

Throughout these challenging years, Oldsmobile faced the task of balancing its performance heritage with modern demands. The design of the 1978-1980 442 reflected this struggle, resulting in a car that was visually distinct from its predecessors yet still recognizably a 442. While it may not have possessed the raw, muscular presence of earlier models, this iteration of the 442 showcased Oldsmobile's ability to adapt its iconic nameplate to changing times.

The downsized 442 may have been smaller in size, but it represented a significant chapter in the model's design evolution. It demonstrated how a legendary performance car could be reimagined for a new era, setting the stage for future revivals and cementing the 442's place in automotive design history.

Section 3.5: Revival and Legacy (1985-1987, 1990-1991)

The Oldsmobile 442's journey didn't end with the downsizing era of the early 1980s. Instead, it experienced a revival that would cement its place in automotive history. This resurgence came in two distinct phases, each offering a unique interpretation of the 442's legendary status.

In 1985, Oldsmobile breathed new life into the 442 nameplate, reintroducing it on the G-body platform. This comeback model struck a delicate balance between classic muscle car aesthetics and contemporary 1980s styling. The 1985-1987 442 featured a muscular stance reminiscent of its forebears, with flared wheel arches and a power bulge hood. However, it also incorporated modern design elements such as aerodynamic front and rear fascias and a more upright greenhouse. The result was a car that looked both familiar and fresh to 442 enthusiasts.

One of the most striking features of this revival was the two-tone paint scheme, often featuring a darker lower body color that accentuated the car's athletic profile. This design choice harked back to the contrasting paint schemes of some earlier 442 models while giving the car a distinctly 1980s flair. The interior likewise blended old and new, with supportive bucket seats and a driver-focused cockpit that included more modern instrumentation and materials.

However, the true final chapter of the 442 story was written in 1990-1991. These models, based on the W-body platform, represented a significant departure from the 442's traditional design language. The 1990-1991 442 adopted a sleek, modern look that reflected the times. Gone were the boxy shapes of the past, replaced by smooth, aerodynamic lines that sliced through the air with minimal resistance.

Despite this modern makeover, Oldsmobile designers were careful to incorporate subtle nods to the 442's heritage. The W-40 package, for instance, included hood stripes that echoed the iconic W-30 models of the muscle car era. These stripes, along with discrete 442 badging, served as visual links to the car's storied past.

The integration of advanced electronics in these final models allowed for a more streamlined dashboard design. Gone were the analog gauges of old, replaced by digital readouts that provided drivers with a wealth of information at a glance. This blend of high-tech features with performance car heritage created a unique driving experience that was both nostalgic and forward-looking.

Color choices for these last 442s were more subdued than their muscle car ancestors, reflecting changing consumer preferences. Metallic grays, deep reds, and midnight blues replaced the bright, attention-grabbing hues of the 1960s and early 1970s. These colors gave the final 442s a more sophisticated, upscale appearance that aligned with Oldsmobile's premium brand positioning.

The most significant aspect of the 442's design in its twilight years was its influence on other Oldsmobile models and the broader automotive industry. Elements of the 442's aggressive stance and performance-oriented design cues can be seen in later Oldsmobile performance models, such as the Cutlass Calais 442 and even the Aurora. The 442's legacy of blending performance with style continued to inspire designers long after the last model rolled off the assembly line.

As we reflect on the design evolution of the Oldsmobile 442, from its muscular beginnings to its sleek final form, we see more than just the progression of a single model. The 442's journey through the decades serves as a mirror to the changing tastes, technologies, and cultural shifts of American society. Each iteration of the 442, whether it was the raw power of the 1960s models or the high-tech sophistication of the 1990s versions, captured the spirit of its era while maintaining a connection to its rich heritage.

The Oldsmobile 442's design story is one of adaptation and resilience. Through changing market demands, stricter regulations, and evolving consumer preferences, the 442 managed to maintain its identity as a performance icon. Its ability to grow while honoring its roots is a testament to the skill and vision of Oldsmobile's designers and engineers. Even in its final years, the 442 continued to turn heads and quicken pulses, proving that great design, like excellent performance, is timeless.

Section 3.6: Design Impact and Cultural Significance

The Oldsmobile 442's design evolution over its production years left an indelible mark on American automotive culture and design philosophy. As we've explored the various styling changes through the decades, it's crucial to understand the broader impact and cultural significance of these design choices.

The 442's design journey mirrored the changing tastes and values of American society. In its early years, the bold, muscular styling of the 442 reflected the optimism and power of 1960s America. The aggressive lines, wide stance, and attention-grabbing colors spoke to a nation that valued strength and individuality. As one automotive historian noted, "The 442's design was a rolling sculpture of American confidence."

As we moved into the 1970s, the 442's design adaptations in response to safety regulations and fuel economy concerns paralleled the nation's growing awareness of environmental and safety issues. The softer lines and more luxurious interiors of the mid-1970s models reflected a shift in consumer preferences towards comfort and practicality without entirely sacrificing performance aesthetics.

The 442's design also played a significant role in shaping the entire muscle car genre. Features like the distinctive hood scoops, bold graphics, and performance-oriented interiors set standards that other manufacturers often emulated. The iconic "442" badging became a symbol of performance that resonated beyond Oldsmobile enthusiasts, entering the broader lexicon of automotive culture.

Interestingly, the 442's design influenced fashion and popular culture as well. Its distinctive color schemes and graphics inspired everything from clothing designs to album covers. The car's appearances in films and television shows of the era further cemented its status as a cultural icon, with its sleek lines and powerful presence making it a natural choice for Hollywood productions seeking to convey speed, rebellion, or American muscle.

The later iterations of the 442, particularly the 1980s models, demonstrated how a classic design language could be adapted for a new era. By incorporating retro elements into modern designs, Oldsmobile created a bridge between nostalgia and contemporary styling, influencing the retro-modern design trend that would become popular across the automotive industry in subsequent decades.

From a technical standpoint, the 442's design evolution also reflects advancements in automotive engineering and manufacturing. The transition from hand-drawn designs to computer-aided drafting and modeling enabled the creation of more precise and complex shapes, particularly evident in the aerodynamic considerations of later models.

The 442's interior design changes over the years showcased the evolution of ergonomics and driver-focused design in performance vehicles. From the driver-centric cockpit layouts to the integration of advanced electronics in later models, the 442 consistently prioritized the connection between driver and machine, setting benchmarks for performance car interiors.

Perhaps most importantly, the 442's design told a story of American ingenuity and adaptability. Through changing regulations, market demands, and technological advancements, the 442's designers consistently found ways to create visually striking and emotionally resonant vehicles. This adaptability in design mirrors the broader American automotive industry's ability to evolve and innovate.

As we reflect on the design legacy of the Oldsmobile 442, it's clear that its impact extends far beyond the realm of automotive enthusiasts. It stands as a testament to the power of design to capture the spirit of an era, influence culture, and create lasting emotional connections. The 442's journey from muscular powerhouse to sophisticated performer, all while maintaining its distinctive character, demonstrates the enduring power of thoughtful, evolving design in the automotive world.

Section 3.7: The 442's Design Legacy

The Oldsmobile 442's design evolution left an indelible mark on automotive history, influencing not only other Oldsmobile models but also the broader automotive industry. Throughout its production run, the 442 consistently pushed the boundaries of performance car aesthetics, setting trends and inspiring competitors.

The impact of the 442's design is evident in subsequent Oldsmobile performance models. The aggressive stance, muscular fender flares, and bold front-end styling that defined the 442 in its prime became hallmarks of Oldsmobile's performance-oriented offerings. For instance, the Oldsmobile Cutlass Calais 442 of the late 1980s, while a far cry from the original muscle car, still incorporated design cues that harkened back to its legendary namesake. The distinctive hood bulges and sporty graphics were clear nods to the 442's heritage.

Beyond Oldsmobile, the 442's influence extended to the entire muscle car segment. Its perfect blend of aggression and sophistication inspired other manufacturers to strike a similar balance in their designs. The concept of a performance package that transformed a mainstream model into a street warrior, as seen in the early 442s, became a blueprint for many other muscle cars of the era.

The 442's design legacy also extends to its ability to adapt and evolve with changing times. From the bold, chrome-laden aesthetics of the 1960s to the more subdued, aerodynamic forms of the 1980s, the 442 demonstrated how a performance car could remain relevant by embracing new design paradigms while maintaining its core identity. This adaptability has become a valuable lesson for modern performance car designers, who must strike a balance between heritage and contemporary aesthetics, as well as meet regulatory requirements.

In the collector car world, the 442's design continues to be celebrated. Restoration projects often aim to recapture the car's original aesthetic appeal, with enthusiasts paying particular attention to period-correct details. The timeless appeal of a well-preserved or accurately restored 442 speaks volumes about the enduring quality of its design.

Moreover, the 442's design legacy can be seen in the modern resurgence of muscle car-inspired vehicles. While Oldsmobile is no longer producing vehicles, the spirit of the 442 lives on in contemporary performance cars that blend retro design cues with modern technology. The bold grilles, aggressive stances, and powerful profiles of today's muscle cars owe a debt to trailblazers like the 442.

Lastly, the 442's design legacy is preserved in popular culture. Its distinctive look has been featured in countless movies, TV shows, and video games, cementing its status as an icon of American automotive design. This cultural presence ensures that the 442's influence continues to inspire new generations of car enthusiasts and designers.

In conclusion, the Oldsmobile 442's design legacy is a testament to the power of thoughtful, evolutionary automotive styling. From its inception as a performance package to its final iterations, the 442 consistently delivered designs that were both of their time and timeless. Its influence on Oldsmobile, the muscle car segment, and the broader automotive industry ensures that the 442 will be remembered not just as a high-performance machine, but as a rolling work of American automotive art.

High Octane Heritage: *Celebrating the Oldsmobile 442*

Chapter 4: Performance Powerhouse: The 442's Engine and Drivetrain

Section 4.1: The Birth of the 442 Engine

The story of the Oldsmobile 442's legendary powerplant begins with a bang, quite literally. In 1964, when Oldsmobile first introduced the 442 option package for the Cutlass, it was clear they weren't just dipping their toes into the muscle car waters; they were diving in headfirst.

At the heart of this new performance beast was the original 330-cubic-inch V8 engine. This wasn't just any run-of-the-mill V8; it was a high-compression, high-output version that set the stage for the 442's reputation as a force to be reckoned with on the streets and strips of America. The 330 V8 was no slouch, producing a respectable 310 horsepower and 355 lb-ft of torque, numbers that raised eyebrows and quickened pulses in equal measure.

However, Oldsmobile engineers knew they were in an arms race. The muscle car era was heating up, and standing still meant falling behind. In 1965, they made a move that would define the 442 for years to come: the introduction of the legendary 400 cubic inch V8. This wasn't just a bump in displacement; it was a quantum leap in performance and a statement of intent from Oldsmobile.

The 400 cubic inch V8 wasn't just bigger; it was better in every way. With 345 horsepower and a stump-pulling 440 lb-ft of torque, it transformed the 442 from a strong contender into a bona fide muscle car icon. This engine would become synonymous with the 442, powering tire-shredding performance that could hold its own against any challenger.

At this point, we should address the elephant in the room, or rather, the numbers on the badge. The "4-4-2" designation has been the subject of much debate and speculation over the years. Initially, it stood for a four-barrel carburetor, four-speed manual transmission, and dual exhausts. However, as the car evolved, so did the meaning. Oldsmobile later reinterpreted it as 400 cubic inches, a four-barrel carburetor, and dual exhausts. Regardless of the specifics, one thing was clear: 442 meant performance.

When compared to its contemporaries, the 442's engine stood tall. While the Pontiac GTO might have kick-started the muscle car craze, and the Ford Mustang captured the public's imagination, the 442's powerplant was a marvel of engineering that could go toe-to-toe with any rival. It offered a combination of raw power, reliability, and refinement that was hard to beat.

The initial power outputs and performance figures of the 442 were nothing short of impressive. With the 400 cubic inch V8, the 442 could rocket from 0-60 mph in just over 6 seconds and clear the quarter-mile in the mid-14 second range. These were numbers that put it squarely in the upper echelon of performance cars of the era.

But more important than the raw numbers was the character of the engine. The 442's V8 was known for its broad power band, delivering strong acceleration throughout the rev range. It wasn't just about peak horsepower; it was about usable, real-world performance that made the 442 a joy to drive on the street and a terror at the drag strip.

The birth of the 442 engine marked the beginning of a performance legacy that would span decades. It set the stage for future innovations and cemented Oldsmobile's place in the muscle car pantheon. As we'll see in the following sections, this was just the beginning of the 442s

Section 4.2: Evolution of the 442 Engine

The Oldsmobile 442's engine underwent a remarkable evolution throughout its production run, continuously pushing the boundaries of performance and adapting to changing times. This journey began with the introduction of the legendary 455 cubic inch V8 engine in 1970, a powerplant that would become synonymous with the 442's muscular reputation.

The 455 V8 was a game-changer for the 442, offering a significant boost in both horsepower and torque compared to its predecessors. When it first appeared, this massive engine produced an impressive 365 horsepower and a staggering 500 lb-ft of torque, figures that put the 442 at the forefront of the muscle car arms race. Over the years, Oldsmobile engineers continued to refine and enhance the 455, squeezing out even more performance. By 1970, the W-30 package bumped the output to 370 horsepower, solidifying the 442's position as a top contender in the muscle car world.

However, the 442's evolution wasn't just about raw power. As the 1970s progressed, the automotive landscape underwent a dramatic shift. Stricter emission standards and rising fuel costs posed significant challenges for high-performance engines. Oldsmobile, like other manufacturers, had to adapt quickly. This led to a series of modifications and innovations aimed at maintaining performance while meeting new regulatory requirements.

One of the most significant changes occurred in 1972, when the industry transitioned from gross to net horsepower ratings. This change, combined with lower compression ratios to accommodate unleaded fuel, resulted in a noticeable drop in official power figures. The 455 V8, which had once boasted 365 horsepower, was now rated at 300 horsepower net. Despite this apparent reduction, the 442's real-world performance remained impressive, thanks to Oldsmobile's engineering prowess in optimizing the engine for the new standards.

Throughout this period of transition, Oldsmobile continued to offer special high-performance engine variants to satisfy enthusiasts. The W-30 package remained available, featuring a host of performance upgrades including a hotter cam, revised cylinder heads, and a specialized air induction system. These enhancements allowed the 442 to maintain its performance edge even as regulations tightened.

As the 1970s wore on, the muscle car era began to wane, and the 442 engine lineup reflected this shift. In 1975, the mighty 455 V8 was relegated to an option, with a smaller 350 cubic inch V8 becoming the standard powerplant. This downsizing trend continued, mirroring the broader industry move towards more fuel-efficient vehicles.

The final years of the original 442 saw further engine downsizing and a focus on efficiency over raw power. By 1980, the largest available engine was a 305 cubic inch V8, a far cry from the massive 455 of a decade earlier. Despite this reduction in displacement, Oldsmobile continued to innovate, introducing technologies like electronic fuel injection to improve performance and efficiency.

The evolution of the 442 engine from its inception to its final years tells a story of American automotive engineering at its finest. It showcases Oldsmobile's ability to adapt to changing times while still striving to deliver the performance that 442 enthusiasts craved. This journey from the thunderous 455 V8 to the more modest but technologically advanced engines of the later years reflects not just the history of the 442 but the broader narrative of the American muscle car era. It's a testament to Oldsmobile's commitment to performance, even in the face of significant challenges, and helps explain why the 442 remains an icon of the muscle car world to this day.

Section 4.3: Innovative Engine Technologies

The Oldsmobile 442's engine wasn't just about raw power; it was a showcase of innovative technologies that set it apart from its competitors. Oldsmobile engineers were constantly pushing the boundaries of engine design, introducing advancements that would not only boost performance but also improve reliability and efficiency.

One of the most significant innovations was Oldsmobile's advanced valve train designs. The 442's engines featured hydraulic lifters, which were self-adjusting and required less maintenance than solid lifters. This design not only improved reliability but also allowed for quieter engine operation. As the 442 evolved, Oldsmobile introduced more advanced camshaft profiles that optimized valve timing for better performance across a wider RPM range.

The carburetion systems of the 442 also saw significant advancements over the years. Early models featured a 4-barrel carburetor, which was advanced for its time. However, the introduction of the Rochester Quadrajet carburetor in later models was a game-changer. The Quadrajet combined the fuel economy of a 2-barrel carburetor at low speeds with the performance of a 4-barrel at high speeds. This innovative design allowed the 442 to be both powerful and relatively fuel-efficient for its class.

One of the most iconic innovations was the W-30 forced air induction system. This factory option included special air intakes mounted outside the grille, which fed cool, dense air directly to the carburetor. The system also featured a fiberglass hood with functional air scoops. This setup significantly increased the engine's breathing capacity, resulting in improved horsepower and torque. The distinctive look of the W-30 package became a hallmark of high-performance 442s.

Ignition system advancements played a crucial role in the evolution of the 442's performance. Oldsmobile introduced transistorized ignition systems, which provided a hotter, more consistent spark. This improvement led to better combustion, increased power, and improved fuel economy. Later models featured electronic ignition systems, further enhancing reliability and reducing maintenance requirements.

Cooling system improvements were essential to support the high-performance applications of the 442's engines. Oldsmobile engineers developed more efficient radiators and water pumps to handle the increased heat generated by higher horsepower engines. They also introduced improved coolant formulations to protect the engine internals better. For high-performance models like the W-30, additional cooling measures such as an aluminum intake manifold and a HD cooling fan were employed to ensure reliability under extreme conditions.

These innovative engine technologies weren't just about numbers on a spec sheet; they translated into real-world performance that drivers could feel and appreciate. The 442's engine was responsive, powerful, and surprisingly versatile. It could cruise comfortably on the highway or tear up a drag strip with equal aplomb. The smooth power delivery and broad torque curve were a testament to the advanced engineering that went into every aspect of the engine's design.

Moreover, these innovations had a lasting impact on the automotive industry. Many of the technologies pioneered or refined in the 442's engine found their way into other General Motors vehicles and influenced competitor designs. The legacy of Oldsmobile's innovative spirit lived on long after the 442 ceased production, continuing to inspire automotive engineers and enthusiasts alike.

In the end, it was this commitment to innovation that helped the 442 stand out in the crowded muscle car market. While other manufacturers may have focused solely on displacement and raw horsepower, Oldsmobile took a more holistic approach, developing technologies that improved every aspect of the engine's performance. This forward-thinking philosophy ensured that the 442's engine remained competitive throughout its production run, cementing its place in automotive history.

Section 4.4: Transmission and Drivetrain

The Oldsmobile 442's impressive engine performance would have been for naught without a robust transmission and drivetrain to harness and deliver its power to the pavement. Oldsmobile engineers recognized this crucial connection and developed a range of transmission and drivetrain options that evolved in tandem with the 442's increasingly powerful engines.

In the early years of the 442, manual transmissions were the preferred choice for performance enthusiasts. The standard offering was a three-speed manual, which provided a direct connection between the driver and the powertrain. However, it was the optional four-speed manual that truly captured the hearts of gearheads. This transmission, often sourced from Muncie, offered closer gear ratios and a higher overall gear range, allowing drivers to keep the engine in its power band more effectively. The four-speed manual became synonymous with the muscle car era, providing the tactile engagement and control that performance drivers craved.

For those who preferred a more relaxed driving experience or wanted the convenience of an automatic, Oldsmobile offered its renowned Hydra-Matic transmission. This three-speed automatic was continuously refined throughout the 442's production run to handle increasing torque outputs and provide quicker shifts. Later models featured a Turbo Hydra-Matic 400, which became legendary for its durability and smooth operation. This transmission was so well-regarded that it was used by other GM divisions and even by competitors like Rolls-Royce.

The rear axle of the 442 was another area where Oldsmobile offered performance-minded options. A variety of gear ratios were available, allowing buyers to tailor their car's performance characteristics to their preferences. Lower numerical ratios provided better acceleration for drag racing enthusiasts, while higher ratios offered improved highway cruising and fuel economy. Many 442s were equipped with limited-slip differentials, which improved traction by transferring power to the wheel with the most grip. This feature was handy for launching the car quickly from a stop and for maintaining control in adverse weather conditions.

Driveshaft technology also saw advancements during the 442's lifetime. Early models used traditional steel driveshafts, but as engine outputs increased, so did the need for stronger components. Oldsmobile introduced larger diameter driveshafts and upgraded U-joints to handle the increased torque. Some high-performance models even featured aluminum driveshafts, which reduced rotational mass and improved acceleration.

The clutch system in the manual transmission 442s underwent significant evolution to cope with the increasing torque of the engine. Early models used single-plate clutches, but as power outputs grew, Oldsmobile introduced heavy-duty and multi-plate clutch systems. These upgrades ensured that the clutch could handle the engine's power without slipping, while still providing a manageable pedal feel for daily driving.

Throughout its production run, the 442's transmission and drivetrain components were continually refined to strike a balance between performance and durability. Oldsmobile engineers worked tirelessly to ensure that every part of the drivetrain could withstand the stress of high-performance driving while still providing a smooth and enjoyable experience for everyday use.

The transmission and drivetrain options offered in the 442 were a key part of its appeal to performance enthusiasts. Whether a buyer chose a manual for maximum control or an automatic for convenience, they could be confident that their 442 was equipped to put its impressive power to the ground effectively. This attention to the entire powertrain package helped cement the 442's reputation as a well-rounded and capable muscle car, ready to perform on the street or the strip.

Section 4.5: Performance Tuning and Modifications

The Oldsmobile 442's robust engine and drivetrain provided an excellent foundation for performance enthusiasts to build upon. Throughout its production run, the 442 offered a range of factory performance packages, dealer-installed options, and aftermarket modifications that allowed owners to tailor their vehicles to their specific performance desires.

At the forefront of factory performance enhancements was the legendary W-30 package. Introduced in 1966, the W-30 option transformed the already potent 442 into a true street warrior. This package included a more aggressive camshaft, special cylinder heads, a performance-tuned carburetor, and the innovative forced air induction system. The W-30 not only boosted horsepower but also provided a distinctive look with its fiberglass hood and functional ram air scoops. As emissions regulations tightened in the early 1970s, the W-30 package evolved, focusing on maintaining performance while meeting new standards.

For those seeking additional performance without opting for the complete W-30 treatment, Oldsmobile offered the L75 option. This package included a higher-output engine with improved components such as a hotter camshaft and revised cylinder heads, providing a noticeable boost in power without the expense of the complete W-30 package.

Oldsmobile dealerships played a crucial role in the 442's performance story, offering a range of dealer-installed options and upgrades. These could include anything from high-performance exhaust systems and cold air intakes to upgraded suspension components and rear axle ratio changes. Dealer-installed options enabled buyers to customize their 442s to their exact specifications while maintaining the factory warranty, a significant advantage in an era when reliability was a concern with heavily modified vehicles.

The aftermarket industry quickly recognized the potential of the 442 platform, developing a wide array of performance parts and accessories. Popular modifications included high-rise intake manifolds, larger carburetors, and more aggressive camshafts. Many enthusiasts also opted for aftermarket headers and free-flowing exhaust systems to unleash more power from their engines. Suspension upgrades, such as stiffer springs and performance shocks, were common additions to improve handling and launch characteristics for drag racing.

However, with great power comes great responsibility, and 442 owners soon learned the importance of striking a balance between performance and reliability. While it was tempting to push the engine to its limits, savvy enthusiasts understood the need for strengthened internal components when significantly increasing horsepower. Forged pistons, stronger connecting rods, and upgraded valve springs became common additions for those seeking to extract maximum performance from their 442s while maintaining longevity.

In recent years, the classic 442 has become a popular platform for modern performance upgrades. Fuel injection conversions have become increasingly common, offering improved drivability and fuel efficiency without sacrificing the engine's characteristic muscle car sound and feel. Electronic ignition systems have replaced points-style distributors, providing more consistent spark and easier tuning. For those seeking massive power gains, forced induction in the form of superchargers or turbochargers has become a popular, albeit more extreme, modification.

Advancements in materials and manufacturing techniques have also enabled the development of high-performance camshafts, cylinder heads, and intake manifolds that significantly exceed the capabilities of original equipment. These modern components enable classic 442 engines to produce power levels that would have been unthinkable during the muscle car era, all while maintaining reliability and drivability.

The world of 442 performance tuning and modifications is a testament to the enduring appeal and robust engineering of Oldsmobile's iconic muscle car. From factory hot rod packages to cutting-edge modern upgrades, the 442 continues to captivate enthusiasts with its performance potential. Whether preserving originality with period-correct modifications or pushing the envelope with modern technology, 442 owners have a wealth of options to tailor their vehicles to their personal performance goals, ensuring that the legacy of Oldsmobile's muscle car continues to thrive on streets and tracks across the country.

Section 4.6: Engine and Drivetrain in Competition

The Oldsmobile 442's engine and drivetrain didn't just shine on the streets; they also made a significant impact in various forms of motorsports. The robust powerplant and sturdy drivetrain components proved to be excellent foundations for competitive racing, allowing the 442 to leave its mark on tracks across America.

In the world of NHRA drag racing, the 442's engine underwent extensive modifications to extract maximum performance. Racing teams would often increase compression ratios, install high-lift camshafts, and add larger carburetors to boost horsepower. The sturdy bottom end of the 455 cubic inch V8 made it particularly popular for drag racing applications, as it could withstand the immense stresses of repeated quarter-mile runs. Many racers also opted for the W-30 package as a starting point, taking advantage of its factory-installed performance upgrades.

Circle track and road racing presented different challenges for the 442's powertrain. While drag racing focused on raw straight-line speed, these disciplines required a balance of power, endurance, and reliability. Engines were often detuned slightly to improve longevity, while cooling systems were upgraded to handle the extended periods of high-rpm operation. Transmissions, particularly in road racing applications, were often fitted with closer-ratio gears to keep the engine in its optimal power band throughout a race.

The 442's competition success was not just theoretical; it racked up numerous notable wins and records. In 1968, a 442 piloted by Dick Arons set an NHRA record in the F/Stock class with a quarter-mile time of 14.12 seconds at 101.23 mph. This achievement demonstrated the potential of even lightly modified 442 engines. In road racing, 442-powered cars consistently performed well in SCCA events, often punching above their weight against purpose-built race cars.

Oldsmobile's commitment to racing was evident in its factory-backed programs. The company provided significant support to teams running 442s in various competitions, supplying them with the latest performance parts and engineering expertise. This symbiotic relationship between the racing teams and Oldsmobile's engineers led to rapid advancements in engine and drivetrain technology.

The most significant aspect of the 442's racing endeavors was the wealth of knowledge gained that directly influenced production models. Lessons learned on the track about engine durability, heat management, and power delivery were often incorporated into subsequent street versions of the 442. For instance, the success of forced air induction systems in racing led to the development of the W-30 package for production cars.

The racing pedigree of the 442's engine and drivetrain also contributed significantly to its street credibility. Enthusiasts knew that the same basic architecture powering their street 442s was also capable of winning races on any given weekend. This racing DNA became a crucial part of the 442's appeal, helping to cement its status as a true performance icon. Ultimately, the 442's success in competition served as a testament to the inherent strengths of its engine and drivetrain design. From the drag strip to road courses, the 442 proved that Oldsmobile could build powertrains that were not just powerful but also versatile and durable enough to excel in the demanding world of motorsports.

Section 4.7: Legacy and Influence

The Oldsmobile 442's engine and drivetrain left an indelible mark on automotive history, influencing not only future Oldsmobile designs but the entire muscle car landscape. This powerplant's legacy continues to resonate with enthusiasts and collectors decades after its final production.

The 442 engine's impact on future Oldsmobile designs was significant. Its success paved the way for other high-performance Oldsmobile models, inspiring engineers to push the boundaries of power and efficiency. The lessons learned from the 442's development were applied to subsequent Oldsmobile vehicles, helping the brand maintain its reputation for performance well into the 1980s and beyond.

Moreover, the 442's powertrain influenced competitor engine development across the industry. As rival manufacturers sought to match or exceed the 442's performance, they were compelled to innovate, leading to a golden age of American muscle. The 442 engine's combination of power, reliability, and adaptability set a high bar that competitors strived to reach, ultimately benefiting car enthusiasts with a wide array of powerful options.

Today, original 442 powertrains are highly sought after by collectors and restorers. The value of these engines has skyrocketed, with well-preserved or expertly restored examples commanding premium prices at auctions and in private sales. This demand is a testament to the enduring appeal and respect for the engineering prowess that went into creating these power plants.

The 442 engine has found new life in modern resto-mod projects. Enthusiasts and builders often choose to transplant these iconic engines into custom builds, blending classic muscle with modern technology. These projects showcase the versatility and enduring performance capabilities of the 442 powertrain, proving that its design remains relevant even in today's automotive landscape.

Numerous organizations and events have sprung up to celebrate and preserve the legacy of the 442 powertrain. Car clubs dedicated to Oldsmobile performance vehicles regularly host meetups, shows, and track days where 442 owners can showcase their vehicles and share knowledge. Annual events like the Oldsmobile Nationals provide a platform for enthusiasts to come together, exchange parts and information, and preserve the spirit of the 442.

Technical workshops and restoration seminars focused on the 442 engine are increasingly popular, attracting both seasoned mechanics and younger generations eager to learn about this iconic powertrain. These educational efforts ensure that the knowledge and skills required to maintain and restore 442 engines are passed down, preserving them for future generations to appreciate.

The 442 engine's influence extends beyond the automotive world, inspiring artists, designers, and even musicians. Its distinctive rumble has been featured in countless films and TV shows, cementing its place in popular culture. The engine's aesthetics, from its valve covers to its air cleaner, have inspired industrial design and artwork, showcasing its impact beyond mere mechanical appreciation.

As we look to the future, the legacy of the 442 engine continues to evolve. With the automotive industry shifting towards electrification, the raw, mechanical power of the 442 serves as a reminder of a different era in performance engineering. Yet, its principles of innovation, efficiency, and pushing the boundaries of what's possible continue to inspire today's automotive engineers and enthusiasts alike.

In conclusion, the Oldsmobile 442's engine and drivetrain have left a lasting legacy that extends far beyond their years of production. From influencing future designs and inspiring competition to becoming prized collector's items and the heart of modern custom builds, these powerplants continue to captivate and excite.

As long as some appreciate the artistry of internal combustion and the thrill of American muscle, the legacy of the 442 engine will endure, a testament to Oldsmobile's engineering excellence and the enduring appeal of pure, unadulterated horsepower.

High Octane Heritage: *Celebrating the Oldsmobile 442*

Chapter 5: Handling and Suspension: Balancing Power with Control

Section 5.1: The Foundation of Control

The Oldsmobile 442's legacy as a muscle car icon is not solely built on its raw power, but also on its remarkable ability to harness that power effectively. At the heart of this capability lies a suspension philosophy that sets the 442 apart from many of its contemporaries. Oldsmobile engineers approached the 442's development with a clear vision: to create a performance car that could not only accelerate rapidly in a straight line but also handle corners with confidence and poise.

This approach stood in stark contrast to some other muscle cars of the era, which often prioritized straight-line speed at the expense of handling finesse. The 442, from its inception, was designed to offer a more balanced performance package. This philosophy was evident in the car's very name - while the "4-4-2" initially stood for four-barrel carburetor, four-speed manual transmission, and dual exhausts, it later came to represent the car's well-rounded nature: four-barrel carburetor, four-on-the-floor, and dual exhausts.

In the early years of the 442, from 1964 to 1967, Oldsmobile laid the groundwork for what would become one of the best-handling muscle cars of its time. The initial suspension setup was based on the Cutlass platform but significantly enhanced for performance duty. The front suspension featured heavy-duty coil springs, recalibrated shock absorbers, and a larger diameter anti-roll bar. At the rear, the four-link design, combined with coil springs, provided a good balance of traction and ride comfort.

The GM A-body platform, which underpinned the 442 along with other notable muscle cars like the Pontiac GTO and Chevrolet Chevelle SS, played a crucial role in the car's handling characteristics. This shared platform provided a solid foundation for Oldsmobile engineers to build upon, allowing them to focus on fine-tuning the suspension components for optimal performance. The A-body's inherent strengths, such as its relatively lightweight design and good weight distribution, gave the 442 an advantage in handling compared to some of its heavier rivals.

Speaking of weight distribution, Oldsmobile paid careful attention to this aspect of the 442's design. The car's front-to-rear weight ratio was carefully balanced to provide neutral handling characteristics. This balance was achieved through the strategic placement of components and, in some cases, the use of lightweight materials. The result was a car that felt nimble and responsive, capable of quick direction changes without the pronounced understeer or oversteer that plagued some other muscle cars of the era.

To further enhance the 442's handling prowess, Oldsmobile engineers implemented several chassis reinforcements. The boxed frame rails, a hallmark of the 442, provided increased structural rigidity compared to the open C-channel design used in standard Cutlass models. This added stiffness not only improved handling by reducing chassis flex during hard cornering but also enhanced the car's overall feel and responsiveness.

Additional structural improvements included strengthened body mounts, which helped to isolate the cabin from road vibrations better while also improving handling precision. The use of a thicker front sway bar and the addition of a rear sway bar in later models further contributed to the car's flat cornering attitude and overall stability.

These foundational elements, the balanced suspension philosophy, the advantages of the A-body platform, attention to weight distribution, and structural reinforcements, combined to create a muscle car that was as adept at carving corners as it was at dominating drag strips. This approach not only set the 442 apart from many of its contemporaries but also laid the groundwork for continuous improvements throughout the model's lifespan.

As we delve deeper into the specific components and their evolution in the following sections, it's essential to keep in mind this underlying philosophy of balanced performance. It was this foundation that allowed the Oldsmobile 442 to remain competitive and relevant throughout its production run, earning it a special place in muscle car history as a true driver's car.

Section 5.2: Front Suspension Evolution

The front suspension of the Oldsmobile 442 underwent significant evolution throughout its production run, reflecting the ongoing quest to strike a balance between power and control. Initially, the 442's front suspension was based on the standard Cutlass setup, but it quickly diverged to meet the demands of performance enthusiasts.

In its early years, the 442 employed a conventional double wishbone suspension system, also known as an SLA (short/long arm) configuration. This setup consisted of upper and lower control arms, coil springs, and hydraulic shock absorbers. While effective for everyday driving, this initial design was soon enhanced to cope with the 442's increasing power and performance expectations.

One of the most significant improvements came with the introduction of the anti-roll bar, also known as a sway bar. This simple yet effective component dramatically improved the 442's handling characteristics. The anti-roll bar connected the left and right sides of the suspension, reducing body roll during cornering and enhancing overall stability. As one side of the car tried to roll outward during a turn, the anti-roll bar transferred some of that force to the opposite side, keeping the car flatter and more composed. This addition was particularly noticeable on winding roads and during high-speed maneuvers, where the 442's considerable horsepower could otherwise overwhelm its chassis.

Coil springs, a crucial element of the front suspension, also saw continuous refinement. Oldsmobile engineers experimented with different spring rates and designs to find the optimal balance between ride comfort and handling performance. As the 442 grew more powerful over the years, stiffer springs were often employed to manage the increased weight and power. For instance, the high-performance W-30 package typically includes specially rated springs to enhance handling without overly compromising ride quality.

Shock absorber technology progressed significantly during the 442's lifetime. The transition from standard to heavy-duty shocks marked a notable improvement in the car's ability to maintain composure over rough surfaces and during aggressive driving. These upgraded shocks provided better damping characteristics, allowing the 442 to recover more quickly from bumps and dips in the road. Some later models even offered gas-charged shocks as part of performance packages, further improving handling and stability.

The steering system, while not strictly part of the suspension, played a crucial role in the 442's handling characteristics. Early models used a recirculating ball steering system, which was robust but sometimes criticized for a lack of feel. Over time, Oldsmobile refined the power steering system, adjusting the steering ratio to provide quicker response and better feedback to the driver. These improvements made the 442 more engaging to drive and easier to control at high speeds.

It's worth noting that many of these suspension improvements were not limited to the 442 alone. As Oldsmobile's engineers developed and refined these components, the lessons learned often trickled down to other models in the lineup. However, the 442 typically received these upgrades first and in their most performance-oriented form.

The evolution of the 442's front suspension is a testament to Oldsmobile's commitment to continuous improvement. Each year brought subtle refinements, with occasional leaps forward in technology or design. This ongoing development helped the 442 maintain its reputation as a muscle car that could do more than just go fast in a straight line; it could corner and handle with the best of them.

By the late 1960s and early 1970s, the 442's front suspension had evolved into a sophisticated system that could effectively harness the car's prodigious power. It provided a level of handling and control that set new standards in the muscle car world, contributing significantly to the 442's legendary status among enthusiasts and critics alike.

Section 5.3: Rear Suspension Developments

The rear suspension of the Oldsmobile 442 played a crucial role in harnessing its immense power and delivering it to the pavement. As the model evolved, so did its rear suspension, with each iteration bringing improvements in handling, traction, and overall performance.

Initially, the 442 utilized a four-link rear suspension system, which was a common configuration for muscle cars of the era. This setup consisted of two upper and two lower control arms that located the rear axle, working in conjunction with coil springs and shock absorbers. The four-link design provided a good balance of lateral stability and axle articulation, allowing the 442 to put its power down effectively while maintaining decent ride quality.

One of the most significant advancements in the 442's rear suspension came with the introduction of the anti-spin differential, also known as a limited-slip differential. This innovative feature dramatically improved traction, especially during acceleration and cornering. The anti-spin differential allowed power to be transferred to the wheel with the most grip, reducing wheel spin and enhancing overall control. This was particularly beneficial for drag racing enthusiasts and drivers who frequently encountered challenging road conditions.

The leaf spring design also underwent significant evolution throughout the 442's production run. Early models utilized traditional multi-leaf springs, which provided adequate support but could sometimes result in axle hop during hard acceleration. Oldsmobile engineers continually refined the leaf spring design, experimenting with different spring rates and configurations to improve both ride quality and performance. Later models introduced asymmetrical leaf springs, with the driver's side spring being slightly stiffer to counteract the torque effect during acceleration. This subtle change helped to reduce wheel hop and improve straight-line stability.

Rear shock absorber technology saw considerable advancements as well. The 442 transitioned from standard twin-tube shocks to more sophisticated gas-charged units in later years. A particularly noteworthy improvement was the introduction of staggered shocks. This configuration placed one shock absorber slightly forward of the axle and the other slightly behind it. The staggered placement helped control axle wrap-up during hard acceleration more effectively, further reducing wheel hop and improving traction.

As the 442's power output increased over the years, so did the need for a stronger rear axle. Oldsmobile responded by upgrading its high-performance models to heavy-duty components. These upgrades included thicker axle shafts, stronger differential gears, and more robust housings. The W-30 package, in particular, featured a heavy-duty rear end designed to withstand the increased torque and power of its high-output engine.

One often overlooked aspect of the rear suspension development was the continual refinement of the rear sway bar. While not present on all models, the rear anti-roll bar became increasingly common on performance-oriented 442s. This addition helped to reduce body roll during cornering and improved overall handling balance, making the car more responsive and predictable at the limit.

The evolution of the 442's rear suspension wasn't just about improving straight-line performance. Oldsmobile engineers worked tirelessly to create a well-rounded system that could handle the diverse demands of a high-performance muscle car. They sought to strike a delicate balance between launch capability, cornering prowess, and ride comfort, a challenging task given the car's dual nature as both a street cruiser and a performance machine.

By the later years of production, the 442's rear suspension had evolved into a sophisticated system that could rival many purpose-built sports cars of the era. The combination of a well-tuned four-link setup, advanced differential technology, carefully engineered springs and shocks, and robust components enabled the 442 to effectively put its considerable power to the ground while maintaining admirable handling characteristics.

This continuous development of the rear suspension played a significant role in cementing the 442's reputation as one of the best-handling muscle cars of its time. It demonstrated Oldsmobile's commitment to creating a true performance machine that excelled not just in straight-line speed but in overall driving dynamics. The lessons learned and technologies developed for the 442's rear suspension would go on to influence future Oldsmobile models and contribute to the broader evolution of performance car engineering.

Section 5.4: Tire and Wheel Configurations

The Oldsmobile 442's tire and wheel configurations played a crucial role in its handling and overall performance. As the muscle car era progressed, so did the technology and design of tires and wheels, with the 442 staying at the forefront of these advancements.

In the early years of the 442, the tire and wheel setups were relatively modest by today's standards. The first 442 models typically rode on 14-inch steel wheels wrapped in bias-ply tires. These tires, while typical for the era, had limitations in terms of grip and handling, mainly when tasked with managing the 442's impressive power output. The standard tire size for the 1964 442 was 7.50-14, which provided adequate performance for daily driving but left room for improvement in high-performance situations.

As the 1960s progressed, Oldsmobile recognized the need for wider tires to improve the 442's traction and handling capabilities. This shift towards wider rubber was a trend across the muscle car segment, as manufacturers sought to put more rubber on the road to harness their vehicles' increasing horsepower. By the late 1960s, the 442 was offered with wider F70-14 tires, which provided a noticeably larger contact patch and improved grip in both straight-line acceleration and cornering.

The evolution of wheel designs was equally important in the 442's development. While early models featured simple steel wheels with hubcaps, Oldsmobile soon introduced more stylish options. The Super Stock I and Super Stock II wheels became iconic choices for the 442, offering both improved aesthetics and performance. These styled steel wheels were not only visually appealing but also stronger and often wider than the standard wheels, allowing for the fitment of larger tires.

As the 1970s dawned, the 442 saw further advancements in its tire and wheel offerings. The introduction of 15-inch wheel options allowed for even wider tires, with sizes like G60-15 becoming available. These larger wheel and tire combinations not only enhanced the car's aggressive stance but also provided tangible benefits in terms of handling and traction.

Performance tire options became increasingly important as the muscle car wars heated up. Oldsmobile offered high-performance tire packages that featured improved compounds and tread designs engineered explicitly for enhanced grip and handling. Goodyear Polyglas tires were a popular choice, offering a blend of performance and durability that complemented the 442's powerful drivetrain.

The impact of these tire technology advancements on the 442's handling cannot be overstated. Compared to the original bias-ply tires, the newer designs offered significantly improved cornering stability, reduced sidewall flex, and better overall road-holding capabilities. This allowed drivers to more fully exploit the 442's performance potential, whether on the street or the track.

It's worth noting that the tire and wheel configurations often varied depending on the specific 442 model and optional packages. For instance, the high-performance W-30 package typically included the widest and most advanced tire options available, further enhancing the car's already impressive performance credentials.

As the muscle car era began to wane in the mid-1970s due to changing regulations and market demands, the 442's tire and wheel configurations continued to evolve. While outright performance may have been less of a focus, advancements in tire technology still brought benefits in terms of ride quality, longevity, and all-weather performance.

The journey of the 442's tire and wheel configurations from modest 14-inch steel wheels to wide, high-performance 15-inch options mirrors the car's overall evolution. As the 442 grew more powerful and refined, its tires and wheels kept pace, ensuring that the car's road-holding abilities matched its straight-line performance. This attention to the crucial interface between car and road was a key factor in establishing the 442's reputation as one of the best-handling muscle cars of its era.

Section 5.5: Brake System Improvements

The Oldsmobile 442's journey to becoming a well-rounded muscle car wasn't complete without significant advancements in its braking system. As the car's power and performance increased over the years, so did the need for more effective stopping power. This section explores the evolution of the 442's brake system, from its initial setup to the introduction of high-performance options.

When the 442 first hit the streets, it was equipped with a conventional brake setup typical of the era. The original configuration featured drum brakes on all four wheels. These drums were large for their time, measuring 9.5 inches in diameter at the front and 9 inches at the rear. While adequate for everyday driving, this setup quickly showed its limitations under hard braking or during prolonged performance driving.

The most significant improvement in the 442's braking system came with the introduction of front disc brakes. This transition began in the late 1960s, with disc brakes initially offered as an option before becoming standard equipment on higher-performance models. The move to disc brakes represented a quantum leap in braking performance. Disc brakes provided several advantages over drum brakes, including improved heat dissipation, enhanced fade resistance, and more consistent performance in wet conditions.

The front disc brake setup typically consisted of single-piston calipers clamping onto ventilated rotors. This design allowed for much more effective cooling of the brake components, crucial for maintaining braking performance during high-speed or repeated hard stops. The rear brakes remained drums, but their size and materials were upgraded to complement the front disc brakes' performance.

As the 442 evolved, Oldsmobile engineers paid increasing attention to brake cooling. They implemented various enhancements to reduce brake fade during performance driving. These improvements included ducting to direct airflow to the brakes and the use of high-temperature brake pad materials. Some models featured notable brake cooling scoops or vents in the front valance, further improving the system's heat management capabilities.

The power brake system also saw significant developments throughout the 442's lifespan. Early models used a simple vacuum-assisted power brake booster, which provided adequate assistance for the drum brake system. As the car's performance increased and disc brakes were introduced, the power assist system was upgraded. Later models featured larger diameter boosters and more sophisticated designs, providing better modulation and a firmer pedal feel. This evolution gave drivers more confidence in the car's braking abilities, especially when pushing the limits of its performance.

For those seeking the ultimate in braking performance, Oldsmobile offered special brake packages on specific 442 models. These high-performance setups typically included larger diameter rotors, more aggressive pad compounds, and upgraded brake lines. Some packages even featured multi-piston calipers for improved clamping force and more even pad wear. These performance brake options were particularly popular on W-30 and Hurst/Olds models, where they complemented the cars' enhanced engine performance.

The improvements in the 442's brake system weren't just about raw stopping power. Engineers also focused on enhancing brake feel and modulation. This attention to detail resulted in a braking system that was not only powerful but also predictable and easy to control. It allowed drivers to brake later when entering corners and gave them the confidence to exploit the car's performance capabilities fully.

By the end of its production run, the Oldsmobile 442's braking system had evolved from a relatively basic setup to a sophisticated system capable of matching the car's impressive acceleration and handling. This evolution played a crucial role in cementing the 442's reputation as a well-rounded performance machine, capable of not just straight-line speed but also confident handling and stopping power.

The advancements made in the 442's braking system throughout its lifespan reflect the broader trend in the muscle car era towards more complete performance packages. As these cars became more powerful and faster, the need for equally capable braking systems became apparent. The 442's brake system improvements stand as a testament to Oldsmobile's commitment to creating a muscle car that was as adept at stopping as it was at accelerating.

Section 5.6: Handling Packages and Special Editions

The Oldsmobile 442's reputation for balancing power with control was further enhanced through various handling packages and special editions. These offerings enabled enthusiasts to push the boundaries of performance, resulting in some of the most coveted versions of the iconic muscle car.

The W-30 package stands out as one of the most significant performance upgrades available for the 442. While primarily known for its engine enhancements, the W-30 also included suspension improvements that complemented the increased power output. The package featured heavy-duty springs and shocks, along with a larger front anti-roll bar. These upgrades resulted in improved handling characteristics, enabling drivers to manage the additional horsepower better. W-30-equipped 442s exhibited noticeably reduced body roll and enhanced cornering capabilities, making them formidable performers both on the street and the track.

The collaboration between Oldsmobile and Hurst Performance produced some of the most sought-after 442 variants. The Hurst/Olds models featured unique suspension tuning that set them apart from standard 442s. These special editions often included stiffer springs, specially valved shock absorbers, and larger anti-roll bars. The result was a more aggressive handling profile that appealed to driving enthusiasts. Hurst/Olds models were known for their exceptional balance, providing a level of responsiveness that belied their size and weight.

For those seeking the ultimate in handling performance, Oldsmobile offered the FE2 suspension package. This Forced Air Induction option was designed to maximize the 442's cornering abilities. The FE2 package included a range of upgrades such as stiffer springs, high-performance gas-pressurized shocks, and larger diameter anti-roll bars. These components worked in harmony to reduce body lean, improve steering response, and enhance overall stability. 442s equipped with the FE2 package were praised for their near-neutral handling characteristics, a rarity among muscle cars of the era.

The Rally suspension option provided another avenue for 442 owners to improve their car's handling prowess. This package was designed to offer a balance between performance and comfort, making it ideal for enthusiasts who used their 442s as daily drivers but still wanted enhanced dynamics. The Rally suspension typically included upgraded springs and shocks, along with a front anti-roll bar. While not as aggressive as the FE2 package, the Rally option noticeably improved the 442's handling without sacrificing ride quality.

As the 442 entered its later years of production, Oldsmobile continued to refine its handling characteristics. Late-model improvements included advancements in suspension geometry, the use of more sophisticated materials in components, and fine-tuning of spring and shock rates. These changes helped the 442 remain competitive in an era when handling performance was becoming increasingly important to buyers. The later models were praised for their poised behavior in corners and improved stability at high speeds, demonstrating that the 442 could evolve with the times.

These handling packages and special editions played a crucial role in cementing the 442's reputation as a muscle car that could do more than just accelerate in a straight line. They showcased Oldsmobile's commitment to creating a well-rounded performance machine, capable of delivering thrills in all aspects of driving. The variety of options allowed buyers to tailor their 442's handling characteristics to their preferences, whether they prioritized all-out performance or a balance of comfort and capability.

Today, these specially-equipped 442s are highly prized by collectors and enthusiasts. They serve as a testament to Oldsmobile's engineering prowess and forward-thinking approach to performance car design. The lessons learned from these handling packages and special editions continue to influence modern performance cars, reminding us that true automotive greatness lies in the balance between power and control.

Section 5.7: Legacy and Influence

The Oldsmobile 442's impact on the muscle car world extended far beyond its impressive performance figures. Its unique approach to balancing power with control left an indelible mark on the automotive industry, influencing future designs and setting new standards for handling in high-performance vehicles.

The 442's handling reputation was a key factor in its success and enduring legacy. Unlike many of its contemporaries that focused primarily on straight-line speed, the 442 earned praise for its well-rounded performance capabilities. Period road tests consistently highlighted the car's composed behavior in corners and its ability to handle the power of its robust engine. Car and Driver magazine, in a 1968 review, noted that the 442 "corners flatter than a flounder on a beach at low tide," a testament to its exceptional suspension tuning and overall balance.

The lessons learned from developing the 442's handling characteristics didn't remain isolated to this model alone. Oldsmobile applied this knowledge across its lineup, elevating the driving dynamics of its other vehicles. The brand's full-size models, like the Ninety-Eight and Delta 88, benefited from improved suspension components and tuning inspired by the 442. Even Oldsmobile's more pedestrian offerings, such as the Cutlass line, saw handling improvements that could trace their roots back to the 442's development.

Perhaps more significantly, the 442's approach to handling raised the bar for muscle car performance as a whole. As competitors took notice of the 442's success and positive reception, they began to place greater emphasis on all-around performance rather than focusing solely on straight-line acceleration. This shift led to a general improvement in muscle car handling across the board, with many manufacturers introducing their own handling packages and suspension upgrades to compete with the 442's balanced performance.

The appreciation for the 442's handling capabilities has only grown over time. Modern enthusiasts and collectors often cite the car's well-rounded nature as a key reason for its desirability. Many current owners praise the 442's predictable handling and comfortable ride, noting how well it combines performance with daily drivability. Restorers often go to great lengths to preserve or recreate the original suspension setups, recognizing their contribution to the car's unique character.

The 442's approach to balanced performance continues to resonate in the world of contemporary performance cars. Modern muscle cars like the Chevrolet Camaro SS 1LE and Ford Mustang GT Performance Pack draw clear parallels to the 442's philosophy, offering enhanced handling capabilities to complement their powerful engines. Even high-end sports cars and supercars have embraced this holistic approach to performance, recognizing that true automotive excellence comes from a harmonious balance of power and control.

In essence, the Oldsmobile 442's legacy in terms of handling and suspension goes far beyond its own model run. It helped reshape the muscle car landscape, pushing the entire industry towards a more sophisticated approach to performance. The 442 proved that an actual performance car could be more than just a straight-line beast; it could be a well-rounded machine capable of tackling any driving scenario with confidence and composure.

This lesson continues to influence automotive design and engineering to this day, ensuring that the spirit of the 442 lives on in the balanced performance cars of the modern era, combined to create a muscle car that was as adept at carving corners as it was at dominating drag strips. This approach not only set the 442 apart from many of its contemporaries but also laid the groundwork for continuous improvements throughout the model's lifespan..

Chapter 6: Special Editions and Rare Variants: Collector's Dream Machine

Section 6.1: The W-30 Package: Performance Perfected

The W-30 package stands as a testament to Oldsmobile's commitment to high-performance engineering, transforming the already potent 442 into a true street beast. Introduced in 1966, the W-30 option quickly became the holy grail for Oldsmobile enthusiasts seeking the ultimate in muscle car performance.

At its core, the W-30 package was designed to extract every ounce of power from the 442's robust engine. The centerpiece of this performance upgrade was a fiberglass hood with functional air scoops, a feature that not only looked aggressive but also served a crucial purpose. These scoops fed cool, dense air directly into the engine, increasing its efficiency and power output. The air intake system was further enhanced with an outside air induction setup, which included special air hoses and a sealed air cleaner, ensuring that only the coolest air possible reached the engine.

Under the hood, the W-30 package included a host of performance-enhancing components. The engine received a more aggressive camshaft, revised cylinder heads with larger valves, and a modified carburetor. These improvements, combined with the enhanced air intake system, resulted in a significant boost in horsepower and torque. For example, the 1970 W-30 442 boasted an impressive 370 horsepower and 500 lb-ft of torque, figures that put it at the forefront of the muscle car horsepower wars.

The W-30's performance enhancements weren't limited to the engine bay. The package also included heavy-duty valve springs, a performance-tuned suspension, and a higher-output battery to handle the increased electrical demands. Some years even saw the inclusion of an aluminum differential cover and shaft to reduce weight and improve performance.

The impact of these upgrades on the 442's performance was dramatic. With the W-30 package, the 442's quarter-mile times dropped significantly. Contemporary road tests of W-30-equipped 442s often recorded quarter-mile times in the low 13-second range, with trap speeds exceeding 105 mph. These were impressive figures for a street-legal production car of the era, cementing the W-30 442's reputation as a formidable force both on the street and at the drag strip.

Despite its impressive performance, the W-30 package was a relatively rare option. Production numbers were limited, making W-30-equipped 442s highly sought after both then and now. For instance, in 1970, often considered the peak year for muscle cars, only 262 W-30 convertibles were built. This rarity, combined with the package's performance pedigree, has made W-30 442s some of the most valuable muscle cars on the collector market.

Today, a well-preserved W-30 442 can command prices well into six figures at auction. The combination of raw power, limited production, and the mystique surrounding the W-30 name has created a perfect storm of desirability among collectors. Particularly rare examples, such as W-30 convertibles or those with documented racing history, can fetch even higher prices.

The W-30 package represents more than just a set of performance upgrades; it embodies the spirit of the muscle car era. It showcases Oldsmobile's engineering prowess and its commitment to delivering top-tier performance to enthusiasts. For many, owning a W-30-equipped 442 is the ultimate expression of muscle car collectorship, a trophy that represents the pinnacle of Oldsmobile's performance legacy.

Section 6.2: The Hurst/Olds: A Collaborative Masterpiece

The Hurst/Olds stands as one of the most iconic special editions in muscle car history, born from a groundbreaking partnership between Oldsmobile and Hurst Performance. This collaboration, which began in 1968, resulted in a series of high-performance vehicles that would leave an indelible mark on the automotive world.

The origins of the Hurst/Olds partnership can be traced back to Oldsmobile's desire to create a standout performance model that would capture the imagination of enthusiasts. Hurst Performance, already renowned for its shifters and performance parts, was the perfect partner to help bring this vision to life. The result was a marriage of Oldsmobile's engineering prowess and Hurst's performance expertise.

What truly set the Hurst/Olds apart were its distinctive features. Perhaps most iconic was the signature gold and white paint scheme, which became instantly recognizable on the street and strip alike. Unique badging, special wheels, and, of course, a Hurst shifter complemented this bold color combination. The interior was equally impressive, often featuring special trim and badging that reminded drivers they were behind the wheel of something extraordinary.

But the Hurst/Olds wasn't just about looks. These machines packed serious performance under the hood. The 1969 model, for instance, boasted a massive 455 cubic inch V8 engine, a full year before this powerplant was available in regular 442s. This engine, combined with other performance enhancements, made the Hurst/Olds a formidable force on both the street and the track.

Adding to their allure was the limited nature of their production runs. In the inaugural year of 1968, only 515 Hurst/Olds were produced, instantly making them collector's items. This trend of limited production continued throughout the partnership, ensuring that each Hurst/Olds model remained a rare and coveted prize.

The legacy and influence of the Hurst/Olds on the muscle car world cannot be overstated. It set a new standard for factory-backed specialty muscle cars, inspiring other manufacturers to create their own high-performance collaborations. The Hurst/Olds demonstrated that with the right partnership, automakers could push the boundaries of performance and style beyond what was typically possible within the constraints of regular production models.

Today, the Hurst/Olds remains one of the most sought-after muscle cars among collectors. Their combination of rarity, performance, and distinctive styling ensures that they continue to command premium prices at auctions and private sales. For many enthusiasts, owning a Hurst/Olds represents the pinnacle of muscle car collecting.

The Hurst/Olds collaboration didn't just produce exceptional cars; it created legends that continue to captivate automotive enthusiasts decades after they first rolled off the assembly line. These vehicles stand as a testament to what can be achieved when two innovative companies come together with a shared vision of performance excellence.

Section 6.3: The 4-4-2 W-Machine: Canadian Muscle

While American muscle car enthusiasts were reveling in the power of the W-30 package and the Hurst/Olds, our neighbors to the north were experiencing their own unique twist on Oldsmobile performance. The 4-4-2 W-Machine, a model exclusive to the Canadian market, represented a distinctive chapter in the 442's storied history.

The W-Machine was born out of a collaboration between Oldsmobile and Canadian dealerships, designed to offer a high-performance 442 variant explicitly tailored for the Canadian market. This special edition first appeared in 1970 and continued through 1972, bridging the gap between the standard 442 and the more extreme W-30 package.

What set the W-Machine apart were its unique visual and performance features. The most striking element was its distinctive appearance. W-Machines sported special striping packages not seen on their American counterparts. These stripes, often in contrasting colors, ran along the length of the car, accentuating its muscular profile. The W-Machine also featured unique badging, proudly displaying its special status to those in the know.

Under the hood, the W-Machine packed a serious punch. Despite stricter Canadian emissions standards of the era, Oldsmobile engineers managed to squeeze impressive performance from these machines. The heart of the W-Machine was typically a high-output version of Oldsmobile's robust 455 cubic inch V8 engine. While exact power figures varied by year and configuration, these engines were known for their prodigious torque, making the W-Machine a force to be reckoned with on Canadian streets and highways.

The interior of the W-Machine also received special attention. Many were equipped with the "rally pack" instrumentation, giving drivers a comprehensive view of the engine's vital signs. Comfort wasn't sacrificed for performance either, with many W-Machines featuring plush bucket seats and other luxury touches that set Oldsmobile apart from its more spartan muscle car competitors.

One of the most intriguing aspects of the W-Machine is its rarity. Production numbers were limited, with only a few hundred units produced each year. This scarcity was partly due to its market-specific nature and partly a result of the declining muscle car market as the 1970s progressed. Today, accurately determining the exact number of W-Machines produced is a challenge, adding to the model's mystique.

The W-Machine's limited production and unique status have made it highly prized among both Canadian and American collectors. For Canadian enthusiasts, it represents a piece of their nation's automotive history, a specially tuned muscle car designed with their market in mind. For American collectors, the W-Machine presents an opportunity to own a piece of Oldsmobile history that never officially crossed the border during its production run.

Values for well-preserved W-Machines have steadily climbed over the years, with immaculate examples commanding premium prices. However, their relative obscurity compared to better-known American muscle cars sometimes allows knowledgeable collectors to find hidden gems.

The 4-4-2 W-Machine stands as a testament to Oldsmobile's commitment to performance across borders. It showcases the brand's willingness to create market-specific variants and highlights the often-overlooked Canadian chapter of the muscle car era. For those lucky enough to own or even encounter a W-Machine, it offers a unique glimpse into a fascinating corner of Oldsmobile's performance legacy.

Section 6.4: The 1970 Rallye 350: Yellow Thunder

In the realm of muscle cars, it's not just about raw power; sometimes, it's also about making a statement. The 1970 Oldsmobile

Rallye 350 did exactly that, combining performance with an unforgettable aesthetic that turned heads wherever it went.

The concept behind the Rallye 350 was both simple and ingenious. Oldsmobile aimed to create a high-visibility performance car that would appeal to younger buyers without breaking the bank. This strategy allowed the division to offer a taste of muscle car excitement at a more accessible price point, broadening the appeal of the 442 lineup.

What truly set the Rallye 350 apart was its distinctive Sebring Yellow paint. This wasn't just any yellow; it was a vibrant, almost luminous shade that made the car impossible to ignore on the street or strip. The bright hue covered not just the body, but also the bumpers, wheels, and even the steering wheel, creating a cohesive and eye-catching look. This bold color choice was complemented by contrasting black stripes and decals, further emphasizing the car's performance pedigree.

Despite its lower price compared to the full-fledged 442, the Rallye 350 still offered robust performance. Under the hood lurked a 350 cubic inch V8 engine, rated at 310 horsepower. This powerplant provided plenty of punch, allowing the Rallye 350 to hold its own against many of its muscle car contemporaries. The package also included a performance-tuned suspension, making it as capable in the corners as it was in a straight line.

The Rallye 350 was more than just a pretty face with a decent engine. It came equipped with a host of performance-oriented features, including a functional hood scoop, dual exhaust, and a heavy-duty cooling system. These elements ensured that the car's performance matched its aggressive appearance.

However, like many muscle cars of the era, the Rallye 350's time in the spotlight was brief. It was only offered for the 1970 model year, with a total production run of 3,547 units. This limited availability has contributed significantly to its collectibility today. The fact that it was a one-year-only model, combined with its distinctive appearance, has made the Rallye 350 a highly sought-after piece of muscle car history.

In the current collector market, the Rallye 350 holds a special place. Its unique color scheme and relative rarity have made it a standout at car shows and auctions. While it may not command the astronomical prices of some ultra-rare muscle cars, well-preserved examples of the Rallye 350 are highly prized by collectors and enthusiasts alike.

The Rallye 350 represents a fascinating chapter in the Oldsmobile 442 story. It showcases the brand's ability to create a visually striking and performant vehicle that could appeal to a broader audience. More than just a marketing exercise, the Rallye 350 proved that sometimes, the most memorable muscle cars aren't always the most powerful or the most expensive. Sometimes, all it takes is a brilliant shade of yellow and a well-rounded package to create an icon.

Section 6.5: The 1971 4-4-2 W-30 Convertible: The Ultimate 442

The 1971 Oldsmobile 442 W-30 Convertible stands as a shining testament to the pinnacle of Oldsmobile's muscle car era. This model year represented the last hurrah for high-compression engines before stricter emissions regulations took effect, making it a true collector's dream and a powerful symbol of an era coming to a close.

What sets the 1971 W-30 Convertible apart is its perfect storm of desirable features. It combined the highly sought-after W-30 performance package with the allure of open-top driving, all wrapped in the muscular styling of the early '70s 442. The result was a car that offered an unparalleled driving experience and would go on to become one of the most coveted muscle cars of all time.

At the heart of this beast was Oldsmobile's mighty 455 cubic inch V8 engine. In W-30 trim, this powerplant was rated at an impressive 350 horsepower, making it one of the most powerful drop-tops of its time. The W-30 package included a host of performance upgrades, such as a fiberglass hood with functional air scoops, a heavy-duty cooling system, and a specially tuned suspension. These enhancements not only boosted performance but also gave the car a more aggressive appearance, setting it apart from standard 442s.

What truly makes the 1971 W-30 Convertible special is its extreme rarity. In a year when muscle car sales were already declining due to rising insurance rates and impending emissions regulations, Oldsmobile produced only 110 W-30 Convertibles. This limited production run has made it one of the rarest 442s ever built, and consequently, one of the most valuable.

The scarcity of these cars, combined with their historical significance as the last of the high-compression muscle cars, has led to astronomical values in today's collector market. A well-preserved 1971 442 W-30 Convertible can fetch prices exceeding half a million dollars at auction. In fact, some remarkably pristine examples have been known to approach the million-dollar mark, putting them in the upper echelons of muscle car collectibility.

Owning a 1971 442 W-30 Convertible is not just about possessing a rare and valuable car; it's about owning a piece of automotive history. These cars represent the zenith of the muscle car era, a time when performance was king and engineers were pushing the boundaries of what was possible with internal combustion engines. They serve as a rolling museum piece, showcasing the pinnacle of Oldsmobile's performance engineering before the automotive landscape changed dramatically in the following years.

For collectors and enthusiasts, the 1971 442 W-30 Convertible offers a unique combination of rarity, performance, and historical significance. It's a car that not only turns heads with its striking appearance but also tells a story of a bygone era in American automotive history. Whether on display at a prestigious car show or cruising down a sunlit highway, the 1971 442 W-30 Convertible continues to captivate and inspire, cementing its status as the ultimate 442.

Section 6.6: The 4-4-2 Pace Cars: Track Stars

The Oldsmobile 442's reputation for performance extended beyond the streets and drag strips to one of the most prestigious racing events in America, the Indianapolis 500. The 442 had the honor of serving as the official pace car for this iconic race not once, but twice, in 1970 and 1972. These appearances at the Brickyard further cemented the 442's status as a true performance icon.

In 1970, the 442 made its debut as the Indianapolis 500 pace car. This was a significant moment for Oldsmobile, as it showcased its flagship muscle car on one of the biggest stages in motorsports. The 1970 pace car was a sight to behold, featuring a striking white and gold color scheme that made it stand out on the track. Oldsmobile, recognizing the marketing potential, offered a limited number of pace car replicas to the public. These replicas were faithful reproductions of the actual pace car, complete with special badging, decals, and the distinctive color combination. For collectors and enthusiasts, owning one of these replicas was like having a piece of racing history in their driveway.

The 1970 pace car replicas were based on the already potent 442 but included some additional performance enhancements to ensure they could lead the race field effectively. They were equipped with the top-of-the-line engine options available that year, making them not just visually striking but also formidable performers.

Just two years later, in 1972, the 442 returned to the Indianapolis Motor Speedway, this time in the form of a Hurst/Olds. This appearance was particularly special as it coincided with the return of George Hurst as the official pace car driver. The 1972 Hurst/Olds pace car featured a unique and eye-catching black and gold color scheme, personally designed by George Hurst. This color combination would become iconic in its own right, further distinguishing the Hurst/Olds models in the muscle car world.

Like its 1970 predecessor, the 1972 Hurst/Olds pace car was also offered as a limited edition replica to the public. These replicas faithfully reproduced the distinctive appearance of the pace car, including its special badging and bold color scheme. They also featured the best performance options available in the Oldsmobile lineup at the time, ensuring that owners could experience a taste of the car's track-worthy performance.

Both the 1970 and 1972 pace cars were equipped with the most powerful engine options available to ensure they could lead the race field effectively. This meant that pace car replicas were not just visually distinctive but also among the best-performing 442s of their respective years.

The significance of these pace cars extends far beyond their appearances at the Indianapolis 500. They represent a high point in the 442's history, a time when Oldsmobile's muscle car was recognized as being worthy of leading the pack at one of the world's most famous races. For collectors, the appeal of these pace cars and their replicas is multifaceted. They combine the inherent desirability of the 442 platform with the added cachet of their Indianapolis 500 connection, distinctive appearances, and limited production numbers.

Today, original pace cars and authentic replicas command premium prices in the collector market. Their historical significance, limited numbers, and unique features make them highly sought after by enthusiasts and collectors alike. A well-preserved pace car or replica can be the centerpiece of any muscle car collection, representing not just the pinnacle of Oldsmobile performance but also a unique moment in motorsports history.

The 442 pace cars serve as a powerful reminder of the model's capabilities and its place in automotive history. They embody the speed, power, and excitement that defined the muscle car era, while their appearances at the Indianapolis 500 underscore the respect and admiration the 442 commanded in its heyday. For many, these pace cars represent the ultimate expression of the 442's performance pedigree - true track stars that shine just as brightly today as they did when they first took to the Brickyard.

Section 6.7: Other Notable Special Editions

While the W-30, Hurst/Olds, and Pace Cars often steal the spotlight, the Oldsmobile 442 lineup boasted several other noteworthy special editions that deserve recognition. These lesser-known variants showcase the diversity and ongoing evolution of the 442 model throughout its production run.

The 1969 Cutlass S Holiday Coupe is often an overlooked gem within the 442 family. While not officially badged as a 442, this special edition offered similar performance in a more understated package. It featured the potent 455 cubic inch V8 engine, typically reserved for the 442, but wrapped in the sleeker Cutlass body style. This combination provided enthusiasts with a "sleeper" option, a car with muscle car performance but without the attention-grabbing appearance of a full-fledged 442. Today, these Holiday Coupes are prized by collectors who appreciate their unique blend of style and power.

As the muscle car era began to wane, Oldsmobile continued to produce special editions that captured the spirit of the 442. The 1975 Hurst/Olds Cutlass represents the last of the classic-era Hurst/Olds models. Despite the challenges posed by stricter emissions regulations and the fuel crisis, this model managed to combine luxury with the last vestiges of muscle car performance. It featured a unique white and gold paint scheme, Hurst shifter, and a still-potent 455 V8 engine. While not as powerful as its predecessors, the '75 Hurst/Olds represented a valiant effort to keep the performance flame alive in challenging times.

High Octane Heritage: *Celebrating the Oldsmobile 442*

The Hurst/Olds name made a triumphant return in 1979, bringing back some excitement to the Oldsmobile lineup. This special edition was based on the Cutlass Calais body and featured a 350 cubic inch V8 engine with a 4-barrel carburetor. The exterior was adorned with the classic gold and white paint scheme, along with unique Hurst/Olds badging and a functional hood scoop. While not as powerful as the Hurst/Olds models of the '60s and early '70s, the 1979 edition represented a significant performance upgrade over standard Oldsmobiles of the time and has become increasingly collectible.

To commemorate the 15th anniversary of the 442, Oldsmobile introduced the 1985 442 Celebration Edition. This limited production model was based on the Cutlass Salon and featured special badging, trim, and a more performance-oriented suspension setup. While its 5.0-liter V8 engine was modest by classic 442 standards, the Celebration Edition honored the legacy of the nameplate and offered enhanced performance compared to the standard Cutlass models of the era. Its rarity and historical significance have made it a sought-after model among 442 enthusiasts.

The 1991 442 W-41 represents one of the last special editions to bear the legendary 442 name. This model marked a significant departure from the V8-powered 442s of the past, instead featuring a high-output version of Oldsmobile's innovative Quad 4 engine. The W-41 package included a specially-tuned 190-horsepower version of this 2.3-liter four-cylinder engine, along with a sport-tuned suspension, unique wheels, and subtle exterior enhancements. While it may seem a far cry from the big-block 442s of the '60s, the W-41 showcased Oldsmobile's ability to adapt the 442 concept to changing times and technologies. Its rarity and status as one of the last 442 models have made it an intriguing collectible for Oldsmobile enthusiasts.

These special editions, spanning from the height of the muscle car era to the early '90s, demonstrate the enduring appeal and adaptability of the 442 nameplate. Each model represents

Oldsmobile's efforts to maintain the performance heritage of the 442, even as market demands and regulations evolved.

For collectors and enthusiasts, these lesser-known variants offer unique opportunities to own a piece of 442 history, often at more accessible price points than their more famous counterparts. Whether it's the sleeper status of the '69 Holiday Coupe, the twilight muscle car era represented by the '75 Hurst/Olds, or the modern interpretation seen in the '91 W-41, these special editions add depth and diversity to the rich tapestry of the Oldsmobile 442 legacy.

High Octane Heritage: *Celebrating the Oldsmobile 442*

Chapter 7: The 442 on the Track: Racing Heritage and Achievements

Section 7.1: The Birth of a Racing Icon

The Oldsmobile 442's journey from street to track marked the beginning of a legendary racing career that would shape its identity for decades to come. In the mid-1960s, as muscle car fever gripped America, Oldsmobile executives made a pivotal decision to enter the 442 into competitive racing. This bold move was driven by the realization that success on the track could translate to increased sales and brand prestige.

To prepare the 442 for its racing debut, Oldsmobile engineers worked tirelessly to modify and enhance the already potent package. They focused on improving the engine's power output, strengthening the drivetrain to handle the increased stress of racing conditions, and fine-tuning the suspension for optimal handling on various track surfaces. These early modifications laid the groundwork for what would become a continuous process of refinement and improvement throughout the 442's racing career.

The 442's first major race appearances were met with a mix of excitement and skepticism from the racing community. Some doubted whether a division of General Motors known primarily for comfortable, family-oriented cars could produce a true contender on the track. However, those doubts were quickly dispelled as the 442 began to showcase its capabilities in various racing disciplines.

One of the 442's earliest and most notable racing debuts came at the 1965 Daytona 500, where it served as the pace car. This high-profile appearance put the 442 in the spotlight and demonstrated Oldsmobile's commitment to performance. Soon after, the vehicle began competing in local drag racing events and regional stock car races, gradually building a reputation for speed and reliability.

The racing community's initial reception of the 442 was one of pleasant surprise. Competitors and spectators alike were impressed by the car's robust performance and the dedication of Oldsmobile's racing teams. As the 442 began to rack up victories and strong finishes, it earned respect from even the most skeptical observers.

Perhaps most importantly, these early racing experiences had a profound influence on the development of production models. The lessons learned on the track directly informed improvements made to street versions of the 442. Engineers used data gathered from race performances to enhance engine tuning, strengthen critical components, and improve overall vehicle dynamics.

For example, the introduction of the W-30 package in 1966 was a direct result of racing experience. This high-performance option included a fiberglass hood with functional air scoops, a more aggressive camshaft, and a tuned exhaust system, all elements that had proven beneficial on the track. The W-30 package became highly sought after by performance enthusiasts, bridging the gap between the race-spec 442s and their street-legal counterparts.

As word of the 442's racing prowess spread, it began to attract a new breed of buyer, one who appreciated not just straight-line speed, but also the handling and durability that came from a race-tested pedigree. This shift in perception helped elevate the 442 from being just another muscle car to a genuine performance icon with racing credentials.

The birth of the 442 as a racing icon marked a turning point for Oldsmobile. It proved that the division could compete with the best in high-performance arenas, challenging preconceptions and setting the stage for future success. As the 442 continued to evolve both on and off the track, it carried with it the lessons and triumphs of these early racing experiences, forever cementing its status as a true American racing legend.

Section 7.2: NASCAR and the 442

The Oldsmobile 442's foray into NASCAR racing marked a significant chapter in its motorsports legacy. When Oldsmobile decided to enter the 442 into NASCAR events, it signaled the brand's commitment to proving its performance capabilities on one of America's most popular racing stages.

The 442's NASCAR journey began in the mid-1960s when Oldsmobile saw an opportunity to showcase its muscle car prowess against fierce competition from other manufacturers. The decision to enter NASCAR wasn't just about winning races; it was a strategic move to enhance the 442's reputation and boost sales of the production models.

Several notable drivers piloted the 442 in NASCAR events, bringing both skill and star power to Oldsmobile's racing efforts. One of the most prominent was Bobby Allison, who achieved remarkable success behind the wheel of the 442. Allison's driving prowess, combined with the 442's performance, resulted in several victories and top finishes that put Oldsmobile in the NASCAR spotlight.

The 442's key victories in NASCAR events were not just triumphs for the drivers but also for Oldsmobile's engineering team. One particularly memorable win came at the Daytona 500, where the 442 demonstrated its high-speed stability and power, solidifying its status as a true contender in the world of stock car racing. These victories helped change perceptions of Oldsmobile, transforming it from a maker of comfortable cruisers to a brand capable of producing race-winning machines.

The rigors of NASCAR competition led to numerous technical innovations that eventually found their way into production models. Engineers constantly refined the 442's engine, suspension, and aerodynamics to gain every possible advantage on the track. These improvements, tested under the extreme conditions of NASCAR racing, often translated into enhanced performance and reliability for street-legal 442s.

For instance, the development of more efficient cooling systems for the race cars led to improved heat management in production models. Similarly, aerodynamic tweaks designed to increase stability at high speeds on NASCAR ovals influenced the styling of later 442 editions, combining form with function in a way that appealed to performance enthusiasts.

The 442's participation in NASCAR had a profound impact on Oldsmobile's reputation in the world of stock car racing. It transformed the brand's image, positioning Oldsmobile as a serious player in the high-performance arena. The success on the track translated into increased interest from consumers, many of whom were drawn to the idea of owning a car with a genuine racing pedigree.

Moreover, the 442's NASCAR exploits provided valuable marketing material for Oldsmobile. Advertisements and promotional materials often highlighted the car's racing success, using phrases like "Race on Sunday, Sell on Monday" to capitalize on the connection between track performance and street credibility.

The NASCAR program also fostered a sense of pride among Oldsmobile employees and dealers. The racing success became a rallying point, boosting morale and creating a shared sense of achievement that permeated the entire Oldsmobile organization.

As the years progressed, the lessons learned from NASCAR competition continued to influence the development of the 442 and other Oldsmobile models. The brand's commitment to racing helped maintain its performance edge, even as changing regulations and market conditions presented new challenges.

In retrospect, Oldsmobile's decision to enter the 442 into NASCAR proved to be a pivotal moment in the car's history. It not only showcased the vehicle's performance capabilities but also played a crucial role in shaping the 442's identity as an actual American muscle car. The roar of 442 engines on NASCAR tracks became synonymous with Oldsmobile's commitment to performance, leaving an indelible mark on the brand's legacy in stock car racing.

Section 7.3: Drag Racing Dominance

The Oldsmobile 442's journey into the world of drag racing was nothing short of spectacular, solidifying its reputation as a force to be reckoned with on the quarter-mile. As muscle car mania swept across America in the 1960s, the 442 found its way onto drag strips, where its potent combination of power and traction made it a natural competitor.

The 442's introduction to the drag racing scene was met with immediate enthusiasm. Racers quickly recognized the potential of its robust V8 engine and torque-rich powerband, ideal for launching hard and maintaining speed through the quarter-mile. Early adopters began modifying their 442s, tweaking everything from the carburetion to the rear axle ratios in search of quicker elapsed times.

Among the pantheon of famous 442 drag racers, few names shine as brightly as Joe Mondello's. Known as "Doctor Olds," Mondello's expertise in extracting maximum performance from Oldsmobile engines earned him a legendary status in the drag racing community. His modified 442s consistently outperformed the competition, setting new standards for what these machines could achieve. Another notable figure was Dick Miller, whose innovative approach to engine building helped numerous 442 racers claim victory.

The 442's record-breaking performances at the drag strip were nothing short of astonishing. In the hands of skilled racers and with the proper modifications, these cars were capable of running the quarter-mile in the low 11-second range, a feat that put them in the upper echelons of muscle car performance. One particularly memorable moment came in 1968 when a modified 442 piloted by Joe Schubeck broke into the 10-second barrier, a phenomenal achievement for a car with its roots in a production vehicle.

Recognizing the marketing potential of drag racing success, Oldsmobile implemented factory-supported drag racing programs. These initiatives provided specially prepared 442s to select racers, along with technical support and resources to keep them competitive. The "Dealer Representative" program, in particular, allowed Oldsmobile dealers to sponsor local racers, creating a grassroots network of 442 ambassadors at tracks across the country.

The drag racing success of the 442 had a profound impact on its marketing and sales. Oldsmobile's advertising campaigns frequently touted the car's drag strip prowess, using headlines like "The Winning Combination" and featuring images of 442s thundering down the quarter-mile. This racing pedigree appealed to performance-minded buyers, many of whom were drawn to showrooms by the allure of owning a street-legal version of these drag strip heroes.

Moreover, the lessons learned on the drag strip directly influenced the development of production 442s. High-performance parts tested in racing often found their way into dealer-installed "W" performance packages or factory options. The legendary W-30 package, for instance, included components like the forced-air induction system and high-lift camshaft that were born from drag racing experience.

The 442's drag racing success also spawned a thriving aftermarket industry. Speed shops and performance parts manufacturers developed a wide array of components specifically for the 442, allowing owners to upgrade their cars for both street and strip use. This ecosystem of performance parts further cemented the 442's reputation as a go-to platform for drag racing enthusiasts.

As the muscle car era reached its zenith, the sight of a 442 staging at the Christmas tree became a familiar and fearsome prospect for competitors. The car's consistent performance and adaptability to various classes of drag racing ensured its continued presence on strips across America, even as emissions regulations began to strangle the performance of newer models.

The legacy of the 442 in drag racing extends far beyond mere timeslips and trophies. It represents a golden age of American performance, where showroom stock vehicles could be transformed into quarter-mile terrors with relative ease. The 442's drag racing exploits not only boosted sales and brand prestige but also fostered a community of enthusiasts who continue to celebrate and race these cars today, ensuring that the thunderous roar of a 442 launching hard off the line remains a thrilling spectacle for generations to come.

Section 7.4: Trans-Am Series and Road Racing

The Oldsmobile 442's versatility as a performance machine extended beyond drag strips and oval tracks. Its foray into road racing, particularly in the Trans-Am Series, showcased the car's handling capabilities and further cemented its reputation as a well-rounded muscle car.

Oldsmobile's decision to enter the 442 in the Trans-Am Series was a bold move that signaled the brand's commitment to proving its performance credentials across all racing disciplines. The Trans-Am

Series, known for its highly competitive fields and challenging road courses, provided the perfect stage for the 442 to demonstrate its agility and endurance.

To prepare the 442 for the rigors of road racing, Oldsmobile engineers made significant adaptations to the standard production model. These modifications included upgraded suspension components, enhanced brake systems, and aerodynamic improvements. The engine was also tuned for sustained high-rpm performance, a crucial factor in road racing where maintaining speed through corners was just as important as straight-line acceleration.

One of the most notable performances by the 442 in Trans-Am competition occurred during the 1970 season. At the Mid-Ohio Sports Car Course, a specially prepared 442 driven by Tony DeLorenzo showcased impressive speed and handling, finishing in the top ten against fierce competition from Ford Mustangs, Chevrolet Camaros, and AMC Javelins. This result proved that the 442 could hold its own against purpose-built pony cars on twisting road courses.

When compared to other muscle cars of the era in road racing scenarios, the 442 often surprised competitors and spectators alike. Its balance of power and handling allowed it to outperform some of its more famous rivals on technical circuits where finesse was as crucial as raw horsepower. The car's performance in these events helped dispel the notion that muscle cars were only good for straight-line speed.

The lessons learned from road racing proved invaluable in improving street versions of the 442. Engineers incorporated insights gained from Trans-Am competition into production models, resulting in better-handling cars for consumers. Improvements in suspension geometry, brake heat dissipation, and weight distribution were incorporated into the 442s, enhancing their appeal to buyers who appreciated a muscle car that could do more than just drag race.

For instance, the optional FE2 suspension package offered on later 442 models was a direct result of knowledge gained from road racing. This package included stiffer springs, recalibrated shock absorbers, and larger sway bars, all of which contributed to sharper handling and improved cornering abilities on public roads.

The 442's participation in the Trans-Am Series and other road racing events also had a significant impact on its marketing. Oldsmobile proudly touted the car's racing pedigree in advertisements and brochures, emphasizing its well-rounded performance capabilities. This racing heritage became a key selling point, attracting buyers who wanted a muscle car that offered more than just straight-line thrills.

While the 442 may not have dominated the Trans-Am Series in the same way it did drag racing, its presence in road racing events was crucial in shaping its identity. It proved that Oldsmobile could build a muscle car that was as comfortable carving corners as it was powering down the quarter-mile. This versatility became a hallmark of the 442, distinguishing it from some of its more singularly focused competitors.

The experience gained in Trans-Am racing continued to influence Oldsmobile's approach to performance car design long after the 442's competition days were over. The emphasis on balanced performance and capable handling became part of Oldsmobile's DNA, influencing future models and cementing the 442's legacy as more than just another straight-line muscle car.

In the end, the 442's forays into Trans-Am and road racing added another layer to its rich performance history. It demonstrated the car's adaptability and engineering excellence, proving that Oldsmobile could compete with the best on any type of track. This diverse racing pedigree contributed significantly to the 442's status as one of the most respected and well-rounded muscle cars of its era.

Section 7.5: Grassroots and Amateur Racing

The Oldsmobile 442's impact on the racing world extended far beyond professional circuits, finding a special place in the hearts of grassroots and amateur racers across America. These enthusiasts, driven by passion rather than corporate sponsorships, played a crucial role in cementing the 442's reputation as a formidable performance machine.

At local drag strips and regional race tracks, the 442 became a favorite among weekend warriors. Its robust engine, sturdy chassis, and relatively affordable price point made it an ideal platform for amateur racers looking to make their mark. Many of these enthusiasts found that with some basic modifications, their 442s could outperform more expensive and exotic competitors.

Success stories from local and regional racing events began to circulate, further enhancing the 442's reputation. In small towns and big cities alike, 442 owners were bringing home trophies and setting local track records. One particularly memorable tale comes from a 1968 event in rural Ohio, where a mildly modified 442 defeated a field of purpose-built race cars, shocking spectators and competitors alike. Such David-versus-Goliath victories were not uncommon and helped build a mystique around the 442 that transcended its factory specifications.

Grassroots racing played a pivotal role in building the 442's reputation for reliability and performance. Unlike the carefully controlled environments of professional racing, amateur events often pushed cars to their limits and beyond. The 442's ability to withstand the punishment of repeated hard launches, high-speed runs, and less-than-ideal maintenance schedules spoke volumes about its robust engineering. This reputation for durability made the 442 not just a weekend warrior, but a daily driver that could be relied upon during the work week as well.

As amateur racers pushed their 442s to the limit, they developed a wealth of knowledge about how to extract maximum performance from these machines. Modifications and tuning tips began to circulate within the community, passed along at race meets, car shows, and eventually in enthusiast magazines. Popular upgrades included high-flow carburetors, aftermarket camshafts, and free-flowing exhaust systems. The creativity of these grassroots racers often led to innovations that were later adopted by professional teams and even influenced factory designs.

Most importantly, amateur racing helped foster a tight-knit community of 442 enthusiasts. Race days became social events where owners could share tips, show off their latest modifications, and forge lasting friendships. This sense of community extended beyond the track, with 442 owners forming clubs, organizing cruises, and attending car shows together. The shared experience of racing their beloved Oldsmobiles created a bond that has endured long after the last factory 442 rolled off the assembly line.

The impact of grassroots racing on the 442's legacy cannot be overstated. While professional racing gave the car prestige, it was the army of amateur racers who truly made the 442 a legend in towns and cities across America. They proved that with skill, determination, and a bit of mechanical know-how, the 442 could hold its own against any competitor.

Today, the spirit of grassroots racing lives on in vintage racing events and at revival drag races. Many of the identical 442s that once thundered down local strips in the '60s and '70s are still competing, now lovingly restored and maintained by a new generation of enthusiasts. These events serve as a testament to the enduring legacy of the 442 and the passionate amateurs who helped write its history one quarter-mile at a time.

The grassroots racing scene did more than just showcase the 442's performance capabilities; it created a lasting culture around the car. This culture of enthusiasm, innovation, and camaraderie continues to influence how the 442 is perceived and celebrated today, ensuring its place not just in the annals of automotive history but in the hearts of car lovers everywhere.

Section 7.6: Racing-Inspired Special Editions

The Oldsmobile 442's success on the racetrack didn't just stay confined to the circuit; it found its way onto the streets through a series of exciting special editions. These race-inspired models served as a bridge between the high-octane world of motorsports and everyday driving, allowing enthusiasts to own a piece of the 442's racing legacy.

One of the most notable racing-inspired special editions was the 1970 Oldsmobile 442 W-30. This street-legal beast was a direct result of Oldsmobile's racing experience, featuring a high-output 455 cubic inch V8 engine that produced a whopping 370 horsepower and 500 lb-ft of torque. The W-30 package included a fiberglass hood with functional air scoops, a heavy-duty cooling system, and a performance-tuned suspension, all elements derived from the 442's track-tested configurations.

Another standout was the 1969 Hurst/Olds, a collaboration between Oldsmobile and Hurst Performance. This limited-edition model boasted a unique paint scheme of Cameo White with Firefrost Gold stripes, a design that became instantly recognizable on both street and strip. Under the hood, it packed a modified 455 cubic-inch V8, capable of propelling the car from 0 to 60 mph in just 5.9 seconds, an impressive feat for the era.

The racing DNA in these special editions wasn't just about raw power. Engineers incorporated numerous performance enhancements learned from countless hours on the track. Improved weight distribution, upgraded braking systems, and refined aerodynamics were all part of the package. The 1970 Rallye 350, while not badged as a 442, shared many components with its muscular sibling and showcased how racing know-how could be applied to create a balanced, agile street performer.

These special editions were typically produced in limited numbers, which has significantly impacted their collectibility over the years. The rarity of these models, combined with their direct connection to Oldsmobile's racing program, has made them highly sought after by collectors and enthusiasts alike. A well-preserved W-30 or Hurst/Olds can command premium prices at auctions, often fetching several times the value of a standard 442 from the same year.

Perhaps most importantly, these racing-inspired special editions played a crucial role in keeping the 442's competitive spirit alive long after it had left the professional racing circuit. They served as rolling showcases of Oldsmobile's performance capabilities, allowing the brand to maintain its high-performance image even as changing regulations and market demands began to impact the muscle car era.

For many enthusiasts, these special editions represented the ultimate expression of the 442's potential. They were more than just cars; they were dreams made real, the chance to own a street-legal race car that could dominate at the stoplight grand prix or turn heads at the local cruise night. In this way, the racing-inspired special editions of the 442 ensured that the thrill of the track was never far from reach, preserving the legacy of this iconic muscle car for generations to come.

Section 7.7: The Legacy of the 442 in Motorsports

The Oldsmobile 442's racing legacy is a testament to its prowess on both the street and the track. Throughout its illustrious career, the 442 amassed an impressive array of achievements that solidified its place in motorsports history. From drag strips to NASCAR ovals, the

442 proved time and again that it was more than just a stylish muscle car; it was a true performance machine.

One of the 442's most significant racing accomplishments was its success in NASCAR. The car's powerful engine and robust chassis made it a formidable competitor on the high-speed ovals. Legendary drivers like Bobby Allison and Cale Yarborough piloted the 442 to multiple victories, showcasing its ability to compete with the best Detroit had to offer. These triumphs on the national stage not only brought glory to Oldsmobile but also demonstrated the 442's reliability and performance capabilities to a broad audience.

The 442's racing success translated directly into street credibility, a crucial factor in the muscle car era. As news of its track victories spread, enthusiasts clamored for the chance to own a piece of this winning heritage. Oldsmobile capitalized on this demand by offering special editions and performance packages that closely mirrored their racing counterparts. This direct link between racing and production models created a halo effect, elevating the entire 442 lineup in the eyes of performance-minded buyers.

The influence of the 442's racing program extended far beyond its own model line. Its success on the track inspired future Oldsmobile performance models, setting a high bar for engineering and performance. The lessons learned from pushing the 442 to its limits in competitive environments led to advancements in engine technology, suspension design, and aerodynamics that benefited the entire Oldsmobile brand.

When compared to the racing heritage of other muscle car legends, the 442 holds its own admirably. While cars like the Pontiac GTO and Ford Mustang may have garnered more mainstream attention, the 442's racing pedigree is every bit as impressive. Its versatility across different racing disciplines, from drag racing to road courses, showcases a depth of performance that few other muscle cars could match.

The most enduring impact of the 442's racing history is its effect on the car's collector status. Today, 442s with documented racing history or those built to racing specifications are among the most sought-after by collectors and enthusiasts. These cars represent the pinnacle of 442 performance and serve as tangible links to the model's glorious past on the track.

The 442's racing legacy continues to influence the automotive world even today. At vintage racing events and muscle car shootouts, 442s continue to compete, carrying on a tradition started decades ago. Their presence on the track serves as a living history lesson, reminding us of a time when American muscle cars ruled the racing world and the 442 was at the forefront of that dominance.

In the end, the 442's racing heritage is more than just a list of victories or records. It's a crucial part of the car's DNA, a testament to Oldsmobile's commitment to performance, and a key reason why the 442 remains an icon of the muscle car era. The roar of a 442 engine on a racetrack isn't just noise; it's the sound of automotive history in motion.

High Octane Heritage: *Celebrating the Oldsmobile 442*

High Octane Heritage: *Celebrating the Oldsmobile 442*

Chapter 8: Cultural Icon: The 442's Impact on American Car Culture

Section 8.1: The Birth of an Icon

The Oldsmobile 442's entrance into the muscle car scene was nothing short of spectacular. In 1964, as America's appetite for high-performance vehicles was reaching a fever pitch, Oldsmobile unveiled a car that would redefine the boundaries of power and style. The 442 burst onto the stage not as a standalone model, but as an option package for the Cutlass, instantly capturing the attention of both performance enthusiasts and casual drivers.

Initial public reception was overwhelmingly positive. Car magazines of the era praised the 442's combination of raw power and refined handling, a balance that set it apart from many of its muscle car contemporaries. Road & Track lauded its "surprising agility," while Hot Rod magazine declared it "a force to be reckoned with on both street and strip." This media coverage played a crucial role in cementing the 442's reputation as a serious contender in the muscle car wars.

High Octane Heritage: *Celebrating the Oldsmobile 442*

What truly set the 442 apart from its competitors was its unique blend of performance and sophistication. While other muscle cars of the time often sacrificed comfort and handling for straight-line speed, the 442 offered a more well-rounded package. Its robust engine was complemented by upgraded suspension, brakes, and transmission, creating a car that was as comfortable carving corners as it was dominating drag strips. This comprehensive approach to performance gave the 442 a distinct edge in a crowded market.

The significance of the "442" nomenclature itself became a part of the car's mystique. Initially standing for four-barrel carburetor, four-speed manual transmission, and dual exhausts, the name evolved to represent the car's cubic inch displacement in later years. This alphanumeric designation became a symbol of power and performance, instantly recognizable to automotive enthusiasts.

Oldsmobile's early marketing strategies for the 442 were pivotal in establishing its image. The company cleverly positioned the 442 as the thinking man's muscle car, appealing to those who wanted performance without sacrificing refinement. Advertisements emphasized not just the car's impressive straight-line speed, but also its handling prowess and daily drivability. One memorable ad campaign proclaimed, "This is no ordinary muscle car," highlighting the 442's unique position in the market.

The impact of these marketing efforts was profound. The 442 quickly gained a reputation as a sophisticated yet potent machine, attracting a diverse range of buyers from young hot-rodders to successful professionals. This broad appeal helped establish the 442 not just as a formidable performer but as a cultural icon that represented the best of American automotive engineering.

As the 1960s progressed, the 442 continued to evolve, with each iteration further solidifying its status as a legend of the muscle car era. Its birth and rapid rise to prominence marked the beginning of a legacy that would influence American car culture for decades to come, setting new standards for what a performance car could and should be.

Section 8.2: The 442 in Popular Culture

The Oldsmobile 442's impact on American culture extended far beyond the realm of automotive enthusiasts. Its powerful presence and iconic status made it a natural fit for various forms of media, cementing its place in the broader popular culture landscape.

The Oldsmobile 442, a classic American muscle car, has made its mark not only on the streets but also on the silver screen, becoming a symbol of power, style, and nostalgia in various films. Its appearances in movies often highlight its iconic design and performance, making it a memorable part of cinematic history.

In Out of the Furnace (2013), the 1968 Oldsmobile 442 plays a prominent role throughout the film. Starring Christian Bale and Woody Harrelson, this gritty drama uses the car as a visual representation of the rugged and raw energy that defines the story. The 442's presence adds a layer of authenticity and character to the film, grounding it in a specific era and aesthetic.

The 1970 Oldsmobile 442 takes center stage in the high-octane chase scene of Demolition Man (1993). This sci-fi action film, featuring Sylvester Stallone, uses the muscle car to amplify the adrenaline and intensity of the sequence. The 442's roaring engine and sleek design make it the perfect choice for a thrilling pursuit, leaving a lasting impression on audiences.

In Black Cat Run (1998), Patrick Muldoon's character drives a 1970 Oldsmobile 442 convertible, which becomes an extension of his personality. The convertible's open-top design and powerful performance reflect the character's rebellious and adventurous spirit, making it an integral part of the film's narrative.

High Octane Heritage: *Celebrating the Oldsmobile 442*

The action-thriller Sweet Girl (2021), starring Jason Momoa, also features a 442, further cementing the car's status as a cinematic icon. Its inclusion in this modern film demonstrates the timeless appeal of the Oldsmobile 442, bridging the gap between classic muscle cars and contemporary storytelling.

The 1968 Oldsmobile 442 makes another notable appearance in The Day the Earth Moved (1974). As the main character's vehicle, it serves as a reliable companion in the face of disaster, showcasing the car's durability and strength. Similarly, in The Hitcher (2007), a 1970 Oldsmobile 442 is featured, adding to the film's suspenseful and intense atmosphere. The car's bold presence complements the movie's dark and gripping tone, making it an unforgettable element of the story.

Across decades and genres, the Oldsmobile 442 has proven to be more than just a car; it's a symbol of power, freedom, and cinematic flair. Its appearances in these films highlight its enduring legacy and its ability to captivate audiences, both on and off the screen.

Music, another cornerstone of popular culture, also embraced the 442. The car's association with power and freedom made it a natural subject for songwriters, particularly in the rock and country genres. Artists like Bruce Springsteen and John Mellencamp, known for their Americana-infused lyrics, often referenced muscle cars like the 442 in their songs, using them as symbols of youth, rebellion, and the open road.

The 442's influence played a crucial role in shaping the broader "muscle car" image in American culture. Its combination of performance, style, and accessibility helped define what many Americans came to expect from a high-performance automobile. The 442 embodied the spirit of the muscle car era: raw power, bold design, and a hint of defiance against the status quo.

Celebrity ownership and endorsements further boosted the 442's cultural cachet. Famous personalities from various fields, including actors, musicians, and athletes, were often seen driving 442s. This high-profile ownership not only increased the car's visibility but also associated it with success and the American dream. The sight of a celebrity behind the wheel of a 442 sent a powerful message about the car's desirability and status.

Perhaps most significantly, the 442 had a profound influence on car enthusiast culture. It became a centerpiece of car shows, drag races, and cruise nights across the country. Enthusiasts formed clubs dedicated to the 442, sharing maintenance tips, restoration advice, and simply reveling in their shared passion for this iconic automobile. These communities fostered a sense of belonging and camaraderie among 442 owners, further enhancing the car's cultural significance.

The 442's impact on popular culture was not limited to its heyday in the 1960s and 1970s. Its influence continued to resonate in later decades, with the car often appearing in period films, TV shows, and music videos set in the muscle car era. This ongoing presence in the media helped introduce new generations to the 442, ensuring its place in the pantheon of American automotive icons.

Moreover, the 442's cultural impact extended to the world of collectibles and memorabilia. Scale models, posters, and other 442-themed items became sought-after collectibles, allowing fans to own a piece of the 442 legend even if they couldn't afford the car itself. This merchandising further expanded the 442's cultural footprint, making it a recognizable symbol even to those who might not be car enthusiasts.

In essence, the Oldsmobile 442's presence in popular culture transformed it from merely a high-performance automobile into a symbol of an era. It came to represent not just automotive excellence, but also the freedom, power, and spirit of innovation that characterized America in the 1960s and 1970s. Through its appearances in various media and its influence on car culture, the 442 transcended its role as a mode of transportation to become a true cultural icon, forever etched in the American consciousness.

Section 8.3: The 442 and American Identity

The Oldsmobile 442 was more than just a car; it was a symbol of American values and aspirations. This powerful muscle car embodied the spirit of freedom, innovation, and individualism that has long been associated with the American way of life. The 442's raw power and sleek design perfectly encapsulated the nation's love affair with the automobile and its desire for constant progress and improvement.

In many ways, the 442 became an integral part of the "American Dream" narrative. It represented success and achievement, serving as a tangible goal for many hardworking Americans. Owning a 442 was seen as a sign that one had "made it," a reward for dedication and perseverance. The car's accessibility to the middle class, compared to more exotic sports cars, made it a realistic aspiration for many, further cementing its place in the American psyche.

The 442's popularity varied across different regions of the United States, reflecting the country's diverse cultural landscape. In the Midwest, where Oldsmobile had its roots, the 442 was particularly revered. It was seen as a homegrown hero, a testament to American engineering and manufacturing prowess. In the South and rural areas, the 442's robust performance made it a favorite among those who appreciated raw power and straight-line speed. On the coasts, particularly in California, the 442 found a home in the burgeoning car culture, becoming a staple at drive-ins, burger joints, and impromptu street races.

For the youth of America, the 442 became a symbol of rebellion and independence. Its aggressive stance and thunderous engine note appealed to a generation that was increasingly questioning authority and seeking to define itself. The 442 provided a means of expression, a way to stand out and make a statement. It was not uncommon to see young people customizing their 442s, adding personal touches that reflected their individuality and desire for self-expression.

As American society underwent significant changes in the 1960s and 1970s, the 442 evolved alongside it. The car's development reflected shifting priorities in American culture, from the initial focus on raw performance to later emphasis on safety and fuel efficiency. The 442's ability to adapt to these changing times while maintaining its core identity mirrored the resilience and adaptability of the American people.

The 442 also played a role in shaping gender dynamics in car culture. While traditionally associated with masculinity, the 442's appeal transcended gender lines. Many women embraced the power and freedom the 442 represented, challenging stereotypes and asserting their place in what had often been a male-dominated space.

In essence, the Oldsmobile 442 became a rolling representation of the American spirit. It embodied the nation's love for power, style, and innovation, while also reflecting its complexities and contradictions. The 442 wasn't just a product of American culture; it actively shaped that culture, influencing everything from fashion and music to social interactions and personal aspirations. Even today, decades after its heyday, the 442 remains a powerful symbol of a specific era in American history, evoking nostalgia for a time when the open road represented endless possibilities and the American Dream seemed within reach for anyone with the courage to chase it.

Section 8.4: The 442's Influence on Car Design and Engineering

The Oldsmobile 442's impact on car design and engineering cannot be overstated. This iconic muscle car pushed the boundaries of performance in ways that reverberated throughout the automotive industry, influencing not just its direct competitors but also shaping the future of car design and engineering.

When the 442 burst onto the scene, it immediately set new standards for what a performance car could be. Its powerful engine, advanced suspension, and innovative drivetrain components raised the bar for the entire industry. Other manufacturers found themselves scrambling to keep up with the 442's impressive performance metrics, leading to a wave of innovation across the board.

One of the most significant contributions of the 442 was its role in advancing handling and safety standards. At a time when many muscle cars were known more for their straight-line speed than their cornering abilities, the 442 stood out for its balanced approach to performance. Its advanced suspension system, which included heavy-duty springs, shock absorbers, and stabilizer bars, provided a level of handling that was unprecedented in its class. This focus on overall performance, rather than just raw power, influenced other manufacturers to pay more attention to the handling characteristics of their vehicles.

The 442's influence extended far beyond its immediate competitors. Its success prompted Oldsmobile and its parent company, General Motors, to incorporate many of its design and engineering principles into their broader lineup of vehicles. The lessons learned from the 442's development were applied to everything from family sedans to luxury cars, helping to elevate the performance and handling capabilities of GM's entire range.

One of the most enduring legacies of the 442 was its impact on the aftermarket modification and customization scene. The car's robust design and high-performance components made it a favorite among enthusiasts looking to squeeze even more power and performance out of their vehicles. This led to a boom in the aftermarket parts industry, with countless companies developing performance upgrades specifically for the 442 and similar muscle cars.

The 442's influence on engine design was particularly notable. Its high-output V8 engine became a benchmark for performance, inspiring other manufacturers to push the boundaries of engine technology. The car's innovative W-30 option package, which included a fiberglass hood with functional air scoops, a high-performance camshaft, and a low-restriction air cleaner, became legendary among car enthusiasts and influenced performance package offerings across the industry.

Even in areas like aerodynamics, the 442 left its mark. While not often associated with streamlined design, the later models of the 442 incorporated subtle aerodynamic improvements that helped improve high-speed stability and fuel efficiency. These lessons would prove valuable as the automotive industry began to place greater emphasis on aerodynamics in the decades that followed.

The 442's impact on transmission technology was also significant. The car's robust three-speed automatic and four-speed manual transmissions were designed to handle the immense power of its engine, setting new standards for durability and performance in automotive transmissions. This influenced the development of stronger, more capable transmissions across the industry.

In the realm of braking technology, the 442 also pushed for advancements. As engines became more powerful, the need for improved braking systems became apparent. The 442's heavy-duty brakes, particularly in its performance packages, helped set new standards for stopping power in high-performance vehicles.

Even today, the influence of the 442 can be seen in modern performance cars. Its balanced approach to performance, combining raw power with handling and braking capabilities, remains a guiding principle in the development of contemporary sports and muscle cars. The lessons learned from the 442's development continue to inform automotive engineers and designers, ensuring that its legacy lives on in the DNA of modern high-performance vehicles.

In conclusion, the Oldsmobile 442's influence on car design and engineering was profound and far-reaching. From pushing the boundaries of engine performance to advancing handling and safety standards, the 442 left an indelible mark on the automotive industry. Its impact can still be felt today, in the powerful, well-balanced performance cars that grace our roads and racetracks. The 442 didn't just change the game; it rewrote the rules, leaving a lasting legacy that continues to inspire automotive innovation.

Section 8.5: The 442 in Advertising and Marketing

The Oldsmobile 442's impact on American car culture extended far beyond its performance on the streets and racetracks. Its influence was equally profound in the realm of advertising and marketing, where it helped shape the way muscle cars were presented to the public and, in turn, how Americans perceived these powerful machines.

Oldsmobile's marketing team recognized the unique appeal of the 442 and crafted campaigns that would become iconic in their own right. One of the most memorable was the "Youngmobile Thinking" campaign, which positioned the 442 as the car for a new generation of drivers who craved both style and power. This campaign perfectly captured the zeitgeist of the 1960s, appealing to young baby boomers who were coming of age and had the purchasing power to match their ambitions.

The visual language of 442 advertisements was bold and dynamic, often featuring the car in action shots that emphasized its power and agility. These images weren't just selling a car; they were selling a lifestyle. The 442 was portrayed as the vehicle of choice for the confident, booming, and adventurous American. This approach reflected the broader cultural shifts of the time, as America embraced youth culture and the idea of personal freedom like never before.

As the 442 evolved over the years, so did its marketing strategies. Early ads focused heavily on performance metrics and engineering prowess, appealing to the gearheads and racing enthusiasts. However, as the muscle car market became more competitive and diverse, Oldsmobile adapted its approach. Later campaigns emphasized the 442's balance of luxury and performance, positioning it as a more sophisticated choice in the muscle car segment.

Compared to its competitors, the 442's marketing often took a more nuanced approach. While brands like Pontiac, with its GTO, usually leaned heavily into aggressive, macho imagery, the 442's campaigns maintained a level of refinement that set it apart. This strategy helped broaden the 442's appeal beyond the typical muscle car demographic, attracting buyers who wanted performance without sacrificing comfort or style.

The impact of 442 marketing extended far beyond its immediate sales figures. These campaigns helped solidify the muscle car's place in the American imagination. They created a visual and emotional language for power and performance that would influence automotive advertising for decades to come. Even today, echoes of those iconic 442 ads can be seen in how modern performance cars are marketed.

Perhaps most importantly, the 442's advertising campaigns contributed to the car's status as a cultural icon. They didn't just sell a product; they sold a dream. For many Americans, the 442 represented the freedom of the open road, the thrill of raw power, and the promise of adventure. These ads tapped into deep-seated American values of individualism and the pursuit of happiness, helping to elevate the 442 from mere transportation to a symbol of the American spirit.

The legacy of 442 marketing can still be felt in the collector car market today. Original advertisements are highly prized by enthusiasts, not just as historical artifacts, but as works of art in their own right. They serve as tangible reminders of an era when American optimism and automotive innovation seemed boundless.

In the end, the marketing of the Oldsmobile 442 did more than sell cars. It helped shape the very culture it was a part of, influencing how Americans viewed not just automobiles, but concepts of freedom, success, and identity. The 442's ad campaigns remain a testament to the power of marketing to transcend its immediate goals and become a part of the broader cultural narrative.

Section 8.6: The 442 Community

The Oldsmobile 442 didn't just create a legacy on the streets and racetracks; it fostered a vibrant, passionate community that continues to thrive today. This community, born out of shared admiration for the iconic muscle car, has played a crucial role in preserving the 442's heritage and keeping its spirit alive.

At the heart of this community are the numerous fan clubs and organizations dedicated to the 442. These groups began forming almost as soon as the car hit the streets, with enthusiasts eager to share their experiences, knowledge, and passion. Over the years, these clubs have grown and evolved, ranging from local chapters to national organizations. They serve as hubs for information exchange, parts sourcing, and social connections among 442 owners and admirers.

One of the most visible manifestations of the 442 community is the array of car shows and events centered around this legendary vehicle. From small local gatherings to large-scale national meets, these events provide opportunities for owners to showcase their prized possessions, share restoration tips, and simply revel in the company of fellow enthusiasts. The annual Oldsmobile Nationals, for instance, has become a pilgrimage of sorts for 442 owners, drawing hundreds of pristine examples and thousands of admirers from across the country.

In the digital age, the 442 community has found new ways to connect and expand its reach. Online forums and social media groups dedicated to the 442 have become invaluable resources for both owners and enthusiasts. These platforms serve as virtual garages where members can troubleshoot mechanical issues, share restoration progress, debate the finer points of 442 history, and organize meet-ups. The online community has been particularly crucial in helping new generations of car enthusiasts discover and appreciate the 442, ensuring that its legacy continues well into the future.

One of the most remarkable aspects of the 442 community is its ability to foster intergenerational connections. It's not uncommon to see grandfathers, fathers, and sons working together on a 442 restoration project, passing down not just mechanical skills but also stories and memories associated with the car. This intergenerational aspect has helped keep the 442's history alive and relevant, bridging the gap between those who experienced the muscle car era firsthand and younger enthusiasts discovering it for the first time.

The 442 community plays a vital role in preserving automotive history. Through meticulous restorations, careful documentation, and passionate storytelling, community members ensure that the 442's legacy is not lost to time. This grassroots preservation effort complements the work of professional automotive historians and museums, offering a rich and multifaceted perspective on the 442's place in automotive history.

Moreover, the community's dedication extends beyond just the cars themselves. Many 442 clubs and organizations are involved in charitable activities, organizing fundraisers, and using their events to support various causes. This aspect of the community highlights how a shared passion for cars can translate into positive action and community service.

The 442 community is more than just a group of car enthusiasts; it's a testament to the enduring appeal of the Oldsmobile 442. It's a living, breathing entity that continues to evolve while staying true to the spirit of the original muscle car. Through their dedication, knowledge sharing, and passion, the members of this community ensure that the legacy of the 442 will continue to inspire and excite generations to come. In many ways, the strength and vibrancy of this community is the most enduring testament to the 442's impact on American car culture.

Section 8.7: The 442's Legacy in Modern Car Culture

The Oldsmobile 442's influence continues to reverberate through modern car culture, leaving an indelible mark on the automotive landscape long after its production ceased. This iconic muscle car has shaped the industry in ways that are still evident today, from influencing modern performance cars to inspiring a dedicated community of enthusiasts.

The 442's impact on modern performance cars is undeniable. Many of today's high-performance vehicles owe a debt to the groundbreaking engineering and design principles pioneered by the 442. The emphasis on powerful engines, responsive handling, and aggressive styling that defined the 442 can be seen in contemporary muscle cars and sports sedans. For instance, the Dodge Challenger and Chevrolet Camaro both draw inspiration from their muscle car heritage, with the 442 being a significant part of that legacy.

In the collector car market, the 442 has seen a steady increase in value and desirability. As original examples become rarer, well-preserved or expertly restored 442s command premium prices at auctions and among private collectors. This trend reflects not only the car's historical significance but also its enduring appeal to automotive enthusiasts. The 442's rising value has also made it a sound investment for collectors, further cementing its place in modern car culture.

The restomod culture, which blends classic car aesthetics with modern performance upgrades, has embraced the 442 with open arms. Enthusiasts and professional builders alike are breathing new life into these classic machines, updating them with contemporary engines, suspension systems, and technology while maintaining their iconic looks. These restomod 442s represent a perfect fusion of nostalgia and modern performance, allowing a new generation of drivers to experience the thrill of a muscle car icon with the reliability and comfort of contemporary engineering.

The 442's historical significance has earned it a place of honor in automotive museums and exhibitions nationwide. From the GM Heritage Center to specialized muscle car museums, the 442 is often featured prominently, telling the story of American automotive innovation and cultural impact. These exhibitions not only preserve the 442's legacy but also educate new generations about its importance in automotive history.

Perhaps most importantly, the 442 continues to inspire new generations of car enthusiasts. Young gearheads who may have never seen a 442 on the road in its heyday are now seeking out these cars, drawn to their raw power, classic styling, and the stories they represent. Social media platforms and online forums buzz with discussions about 442 restoration projects, performance upgrades, and appreciation for the model's history. This ongoing interest ensures that the 442's legacy will continue to thrive in the digital age.

The 442's influence extends beyond just car enthusiasts. It has become a symbol of a bygone era in American culture, representing the freedom, power, and innovation of the muscle car age. Even for those who may never own one, the 442 remains an iconic piece of

Americana, its image instantly recognizable and evocative of a specific time and place in history.

As we look to the future of automotive design and culture, the 442's legacy serves as both an inspiration and a benchmark. It reminds us of the impact that a well-designed, powerful, and culturally significant vehicle can have, not just in its own time, but for generations to come. The Oldsmobile 442 may no longer be in production. Still, its spirit lives on in the hearts of enthusiasts, the designs of modern muscle cars, and the ongoing celebration of American automotive heritage.

High Octane Heritage: *Celebrating the Oldsmobile 442*

Chapter 9: Restoration Guide: Bringing a 442 Back to Life

Section 9.1: Assessment and Planning

Before diving into the restoration of your Oldsmobile 442, it's crucial to begin with a thorough assessment and careful planning. This initial stage sets the foundation for a successful restoration project and helps you avoid costly mistakes down the road.

The first step in this process is conducting an initial evaluation of the vehicle's condition. This involves a comprehensive inspection of every aspect of the car, from the body and frame to the engine, drivetrain, and interior. Look for signs of rust, structural damage, missing parts, and previous repairs or modifications. Pay close attention to the car's numbers-matching components, as these can significantly impact the vehicle's value and authenticity.

As you assess the car, it's essential to document its current state meticulously. Take numerous high-quality photographs from various angles, capturing both the overall condition and specific areas of concern. Make detailed notes about your observations, and consider creating diagrams to map out areas that require attention. This documentation will serve as a valuable reference throughout the restoration process. It can also be helpful for insurance purposes or if you decide to sell the car in the future.

With a clear understanding of your 442's condition, the next step is to research the specific model year and specifications. Oldsmobile produced the 442 across several generations, each with its unique features and options. Consult factory literature, old magazines, and enthusiast websites to gather as much information as possible about your particular model. This research will help you make informed decisions about which parts to use and how to approach various aspects of the restoration.

Armed with this knowledge, it's time to create a comprehensive restoration plan and timeline. Break down the project into manageable phases, such as disassembly, body work, mechanical restoration, and reassembly. Estimate the time required for each phase, considering your available time, skills, and resources. Be realistic in your planning, as restorations often take longer than initially anticipated.

Finally, one of the most critical aspects of planning your 442 restoration is budgeting. Restoration projects can quickly become expensive, so it's vital to establish a realistic budget from the outset. Research the costs of parts, materials, and any specialized services you may need, such as machine shop work or upholstery. Don't forget to factor in unexpected expenses, as surprises are common in restoration projects. It's wise to add a contingency of at least 20% to your estimated budget to cover unforeseen costs.

Remember, a well-executed assessment and planning phase can make the difference between a smooth, successful restoration and a project that becomes overwhelming and potentially abandoned. Take the time to thoroughly evaluate your 442, document its condition, research its history, create a detailed plan, and establish a realistic budget. With this solid foundation in place, you'll be well-prepared to embark on the exciting journey of bringing your Oldsmobile 442 back to its former glory.

Section 9.2: Disassembly and Parts Cataloging

The disassembly and parts cataloging phase is a critical step in the restoration process of your Oldsmobile 442. This stage sets the foundation for a successful restoration project, ensuring that every component is accounted for and properly assessed.

Begin the disassembly process with a methodical approach. Start by removing the larger, easily accessible components such as the hood, trunk lid, and doors. As you work your way through the vehicle, use proper disassembly techniques to avoid damaging parts that may be fragile due to age or corrosion. Always refer to the factory service manuals for guidance on the correct disassembly order and any special tools that may be required.

As you remove each part, implement a robust organizing and labeling system. Use plastic bags or containers to store small parts, and label them clearly with the component name and its location on the vehicle. For larger parts, consider using a tagging system with durable labels that can withstand cleaning processes. Take numerous photographs throughout the disassembly process, capturing the location and orientation of parts before removal. These photos will prove invaluable during reassembly.

While disassembling, carefully evaluate each part for restoration, replacement, or fabrication. Some components may be in good condition and only require cleaning and refinishing. Others may be beyond repair and need replacement. For hard-to-find parts, consider having them fabricated by a specialist. Make detailed notes on the condition of each part and your decision for its fate in the restoration process.

Creating a detailed inventory of all components is crucial. Use a spreadsheet or specialized restoration software to catalog every part, noting its condition, any part numbers, and your plans for restoration or replacement. This inventory will serve as your roadmap throughout the restoration process, helping you track progress and manage your budget.

Sourcing original or reproduction parts is often one of the most challenging aspects of a 442 restoration. Start by identifying which parts need replacement and researching your options. Original parts can frequently be found through swap meets, online marketplaces, or specialty Oldsmobile parts suppliers. For parts that are no longer available, high-quality reproductions may be your best option. Always verify the fit and quality of reproduction parts before purchasing, as not all are created equal.

When sourcing parts, consider joining Oldsmobile enthusiast clubs or online forums. These communities can be invaluable resources for locating rare parts and getting advice on the best suppliers. Don't hesitate to network with other 442 owners and restorers – the classic car community is often beneficial and may lead you to parts you thought were impossible to find.

Remember, patience is key during this phase. Rushing through disassembly and parts cataloging can lead to lost components, damaged parts, and headaches down the road. Take your time, stay organized, and document everything thoroughly. This meticulous approach will pay dividends as you progress through your 442 restoration, ensuring that when it comes time to reassemble your classic muscle car, you'll have all the pieces of the puzzle ready to go.

Section 9.3: Body and Frame Restoration

The body and frame restoration is arguably the most visible and transformative aspect of bringing a 442 back to life. This process requires patience, skill, and attention to detail to ensure your classic Oldsmobile not only looks stunning but also maintains its structural integrity.

Addressing rust and structural issues is the first critical step in body restoration. Rust is the enemy of any classic car enthusiast, and the 442 is no exception. Begin by thoroughly inspecting the vehicle for rust spots, paying close attention to common problem areas such as the floor pans, trunk, wheel wells, and lower body panels.

Small surface rust can often be treated with rust converters and body filler, but more severe cases may require cutting out the affected areas and welding in new metal patches. It's crucial to address all rust issues, no matter how small, to prevent future spreading and ensure a long-lasting restoration.

Frame straightening and reinforcement may be necessary if your 442 has suffered significant damage or years of neglect. Start by placing the frame on a frame jig to check for any misalignment. Use hydraulic tools and heat to straighten any bent sections carefully.

If the frame exhibits signs of weakness or has a history of poor repairs, consider reinforcing critical areas with additional steel plates or boxing in sections of the frame to enhance its strength. Remember, a solid frame is essential for both safety and proper body panel fitment.

Body panel repair or replacement is often one of the most time-consuming aspects of restoration. Carefully assess each panel for damage, rust, or previous poor maintenance. Minor dents and dings can usually be repaired using body hammers, dollies, and fillers.

However, severely damaged or rusted panels may need to be replaced entirely. When sourcing replacement panels, try to find original GM parts or high-quality reproductions to ensure proper fit and authenticity. Pay close attention to panel gaps and alignment during this process, as appropriate fitment will significantly impact the final appearance of your 442.

Surface preparation for painting is a crucial step that can make or break the final finish of your restoration. After all body work is complete, thoroughly sand the entire body to create a smooth, even surface. Start with coarse-grit sandpaper to remove any remaining imperfections, then progressively move to finer grits for a glass-like finish. Apply a high-quality primer to the bare metal, ensuring even coverage. Once the primer has cured, block sand the entire body to achieve a perfectly smooth surface ready for paint.

Paint selection and application techniques are where your 442 truly comes back to life. Research the original factory color options for your specific model year to maintain authenticity. If you're aiming for a concours-level restoration, try to source the exact paint code and formulation used by Oldsmobile.

For a more custom look, consider period-correct colors or modern interpretations of classic hues. Regardless of your color choice, invest in high-quality paint and proper application equipment. Apply multiple thin coats, allowing adequate drying time between each layer. Finish with several coats of clear for depth and protection.

Throughout the body and frame restoration process, take your time and don't rush. Proper preparation and attention to detail at this stage will pay dividends in the final appearance and longevity of your restored 442. Remember to document each step with photographs and notes, not only for your own reference but also to showcase the extent of the restoration to future admirers or potential buyers.

By the end of this stage, your once-neglected 442 will be well on its way to turning heads once again, its gleaming body a testament to the craftsmanship and dedication invested in its restoration. With the exterior taking shape, you're now ready to turn your attention to the heart of the beast - the engine and drivetrain.

Section 9.4: Engine and Drivetrain Rebuilding

The heart of any muscle car restoration lies in the engine and drivetrain rebuild. For the Oldsmobile 442, this process is crucial to recapturing the raw power and performance that made this model legendary.

Begin by carefully disassembling the engine, documenting each step with photographs and notes. As you remove components, inspect them meticulously for wear, damage, or signs of previous repairs. Pay close attention to the cylinder walls, crankshaft, and camshaft, as these components often show telltale signs of the engine's overall condition.

Once disassembled, it's time for a thorough cleaning and evaluation. Use a combination of chemical cleaners and mechanical methods to remove years of grime and buildup. This clean slate allows for a more accurate assessment of each part's condition.

With a clear picture of what needs attention, consult with a reputable machine shop for necessary work. This may include boring the cylinders, resurfacing the cylinder heads, or turning the crankshaft. Quality machine work is essential for ensuring optimal performance and longevity in your rebuilt engine.

When it comes to parts replacement, strive for a balance between originality and reliability. While using period-correct parts maintains authenticity, consider upgrading specific components for improved durability. For instance, modern piston rings or valve seals can significantly enhance engine longevity without compromising the car's classic feel.

The engine reassembly process demands patience and precision. Follow factory torque specifications and use assembly lubricants where appropriate. Pay special attention to timing chain installation and valve lash adjustment, as these can significantly impact engine performance.

Once the engine is reassembled, it's time to focus on the transmission. Whether your 442 came equipped with a manual or automatic transmission, a complete overhaul is typically necessary. This involves replacing worn clutch components in manual transmissions or rebuilding the valve body and replacing worn clutch packs in automatics. Don't forget to address the shifter mechanism, ensuring smooth and precise gear changes.

The rear axle and differential also require attention. Inspect the ring and pinion gears for wear, and replace the bearings and seals as necessary. If you're aiming for enhanced performance, consider upgrading to a limited-slip differential for improved traction.

As you near completion of the drivetrain rebuild, it's crucial to tune the engine properly. This involves setting the ignition timing, adjusting the carburetor (or fuel injection system in later models), and ensuring all systems work in harmony. A dynamometer session can be invaluable for fine-tuning and verifying that your 442's powerplant is performing up to its full potential.

Remember, rebuilding the engine and drivetrain is not just about raw power – it's about recreating the unique driving experience of the Oldsmobile 442. Pay attention to details, such as the distinctive sound of the exhaust and the feel of the power delivery. These elements contribute significantly to the car's character and the overall satisfaction of the restoration project.

Lastly, don't overlook the importance of proper break-in procedures for your newly rebuilt engine. Follow manufacturer recommendations for initial oil changes and driving patterns to ensure long-term reliability and performance.

With patience, attention to detail, and a commitment to quality, rebuilding the engine and drivetrain of your Oldsmobile 442 will result in a powertrain that not only looks period-correct but performs like it did when it first rolled off the assembly line – or perhaps even better.

Section 9.5: Interior Restoration

The interior of your Oldsmobile 442 is where you'll spend most of your time, making its restoration crucial for both authenticity and enjoyment. A well-executed interior restoration can transport you back to the golden age of muscle cars every time you slide behind the wheel.

Begin with the upholstery, the most visible aspect of your 442's interior. Depending on the condition, you may need to repair or completely replace the seat covers. For the most authentic look, source reproduction upholstery that matches the original patterns and materials. Pay close attention to the details, such as the correct grain of the vinyl and the proper stitching patterns. If your budget allows, consider hiring a professional upholsterer to ensure a factory-fresh appearance.

The dashboard and instrument panel are the command center of your 442, and their restoration requires patience and precision. Start by carefully removing all components, including gauges, switches, and trim pieces. Clean each item thoroughly, and repair or replace as necessary.

For plastic parts that have become brittle or discolored over time, specialized restoration products can often bring them back to life. When it comes to gauges, decide whether to restore the originals or install modern reproductions. While original gauges maintain authenticity, reproductions often offer improved reliability and accuracy.

Carpet and headliner replacement is next on the agenda. Factory-correct reproduction carpet kits are readily available for the 442, making this a relatively straightforward task. However, proper fitment requires attention to detail.

Take your time to ensure the carpet lies flat and fits snugly around all obstacles. The headliner can be more challenging, often requiring the removal of the windshield for proper installation. If you're not comfortable tackling this yourself, consider seeking professional help to avoid wrinkles or sagging.

Door panels and trim restoration is another crucial aspect of bringing your 442's interior back to life. Carefully remove the door panels, being mindful of any clips or fasteners that may have become brittle over time.

Clean and repair any damaged areas, paying special attention to armrests and map pockets that often show wear. If the panels are beyond repair, high-quality reproductions are available. When reinstalling, ensure all weather stripping is in good condition to prevent water intrusion and wind noise.

Finally, focus on restoring or replacing the steering wheel and controls. The steering wheel is a focal point of the interior and often shows significant wear. If the original wheel is in good condition, it can usually be restored with specialized products designed to rejuvenate hard plastic and vinyl surfaces.

For severely damaged wheels, consider a reproduction that matches the original design. Don't forget to address other controls such as the shifter, pedals, and horn ring, ensuring they all function correctly and look period-correct.

Throughout the interior restoration process, strive for a balance between authenticity and functionality. While maintaining originality is essential, don't shy away from subtle upgrades that can enhance your driving experience without compromising the classic look. For example, adding sound-deadening material during the carpet installation can significantly improve comfort without altering the visual appeal.

Remember, patience is key when restoring your 442's interior. Take your time with each step, and don't rush the process. The result will be a cabin that not only looks stunning but also provides a comfortable and authentic driving experience, allowing you to fully immerse yourself in the muscle car era every time you take your 442 for a spin.

Section 9.6: Electrical System Overhaul

Restoring the electrical system of your Oldsmobile 442 is a critical step in bringing your classic muscle car back to life. A well-functioning electrical system ensures that all components, from the ignition to the lighting, operate smoothly and reliably. In this section, we'll dive into the intricacies of overhauling your 442's electrical system.

Let's start with the wiring harness, the backbone of your car's electrical system. Over time, wiring can become brittle, corroded, or damaged, leading to electrical gremlins that can be frustrating to diagnose and repair. Begin by carefully inspecting the entire wiring harness for signs of wear, fraying, or previous improper repairs. In many cases, it's advisable to replace the whole wiring harness with a new, reproduction unit. This ensures compatibility with your 442's original specifications while providing the reliability of modern materials.

For instance, a restored 1970 442 owner reported persistent electrical issues until they replaced the entire wiring harness. The result was a dramatic improvement in reliability, along with the elimination of intermittent electrical problems that had plagued the car for years.

Next, consider upgrading to modern fuse boxes and relays. While purists might argue for maintaining complete originality, the safety benefits of modern electrical components can't be overstated. Modern fuse boxes offer better circuit protection and are less prone to failure than the original units. Similarly, modern relays can handle higher electrical loads more efficiently, reducing the risk of electrical fires, a concern with any classic car.

When it comes to gauges and switches, you have a decision to make: restore the originals or replace them with reproduction units. Original gauges can often be refurbished by specialists who can clean, recalibrate, and even replace the internal mechanisms while preserving the original face. This approach maintains the authentic look of your 442's dashboard. However, if the originals are beyond repair or missing, high-quality reproductions are available that closely mimic the look and function of the factory units.

The lighting system is another area that deserves close attention. All exterior lights should be checked for proper operation and replaced if necessary. This includes headlights, taillights, turn signals, and any auxiliary lights your 442 may have. Consider upgrading to modern halogen or LED bulbs for improved visibility and safety, but be sure to check local regulations regarding acceptable lighting modifications for classic cars.

Lastly, consider adding period-correct accessories to enhance the functionality and appeal of your 442. This could include items like an AM/FM radio, power windows, or even air conditioning. While these weren't standard on all 442 models, they were available options and can significantly increase the comfort and enjoyment of your restored muscle car.

For example, one 442 restorer added a period-correct AM/FM radio with modern internals, allowing for Bluetooth connectivity while maintaining the original look. This blend of old and new perfectly captured the spirit of the 442 while adding modern convenience.

Throughout the electrical system overhaul, it's crucial to document all changes and keep a detailed wiring diagram. This will be invaluable for future maintenance or troubleshooting. Additionally, always use the correct gauge wire and proper connectors for all electrical work to ensure safety and reliability.

Remember, patience is key when working on your 442's electrical system. Take your time, double-check all connections, and don't hesitate to seek the help of a professional if you're unsure about any aspect of the electrical restoration. With careful attention to detail and a methodical approach, you'll soon have a fully functioning electrical system that will keep your 442 running smoothly for years to come.

Section 9.7: Final Assembly and Detailing

As you approach the final stages of your Oldsmobile 442 restoration, the excitement builds. All your hard work is about to pay off as you bring the various components together to create a cohesive, stunning vehicle. This phase requires patience, attention to detail, and a methodical approach to ensure everything comes together perfectly.

Begin the final assembly process by carefully following your documented disassembly notes and photos. Start with the larger components and work your way down to the more minor details. Install the rebuilt engine and transmission, ensuring all mounts and connections are secure. Next, focus on the suspension components, brakes, and wheels, making sure everything is properly aligned and torqued to specification.

As you reassemble the body panels, take extra care to achieve proper fitment and alignment. This step is crucial for achieving that factory-fresh look. Use shims and adjusters as needed to get the panels, doors, hood, and trunk lid to sit perfectly. Pay close attention to panel gaps, ensuring they are uniform and consistent with factory specifications. This attention to detail will make a significant difference in the final appearance of your 442.

Installing the glass and weatherstripping is a delicate process that requires patience and precision. Start with the windshield, using the correct adhesive and following the manufacturer's instructions carefully. Install the side and rear glass, along with all new weatherstripping, to ensure a watertight seal. Take your time with this step, as proper installation will prevent future issues with leaks and wind noise.

With the exterior taking shape, turn your attention to the interior. Install the restored dashboard, being careful not to scratch or damage the newly refinished surfaces. Connect all electrical components, double-checking your wiring to ensure everything functions correctly. Install the seats, carpeting, headliner, and door panels, taking care to align everything properly for a factory-fresh look.

The final detailing phase is where your 442 truly comes to life. Start with a thorough cleaning of the entire vehicle, inside and out. Use appropriate cleaners for each surface, being mindful of newly painted areas. Apply a high-quality wax or ceramic coating to protect the paint and enhance its shine. Clean and dress all rubber and plastic components to restore their luster.

Inside the car, clean and condition the upholstery, using products specifically designed for the materials in your 442. Polish all chrome and stainless steel trim to a mirror finish. Clean the glass thoroughly, ensuring there are no streaks or residue left behind.

As you complete the final assembly and detailing, take a moment to appreciate the transformation that has occurred. What was once a neglected classic has been reborn through your dedication and hard work. Your Oldsmobile 442 now stands as a testament to the model's enduring appeal and your commitment to preserving automotive history.

Remember, the restoration process doesn't truly end here. Regular maintenance, careful driving, and continued care will ensure your 442 remains in top condition for years to come. As you prepare for that first drive in your fully restored muscle car, take pride in knowing you've not only preserved a piece of automotive history but have also created a rolling work of art that will turn heads and ignite passion wherever it goes.

Chapter 10: Maintenance Mastery: Keeping Your 442 in Prime Condition

Section 10.1: Engine Maintenance

The heart of your Oldsmobile 442's performance lies within its powerful engine. Proper engine maintenance is crucial not only for preserving the car's value but also for ensuring it continues to deliver the exhilarating performance that made the 442 a legend. Let's delve into the essential aspects of engine maintenance that every 442 owner should be aware of.

Regular oil changes are the foundation of good engine health. For most 442 models, it's recommended to change the oil every 3,000 miles or 3 months, whichever comes first. The type of oil you use is just as important as the frequency of changes. For instance, a 1970 442 with the 455 V8 engine performs best with 10W-30 or 10W-40 oil. These multi-grade oils provide excellent protection across a wide range of temperatures, ensuring your engine stays lubricated whether you're cruising on a hot summer day or firing up on a chilly morning.

Next, let's talk about air filters. A clean air filter is crucial for your 442's engine to breathe correctly. A clogged air filter can reduce fuel economy by up to 10% and decrease acceleration, robbing your muscle car of its signature power. Inspect your air filter every 15,000 miles and replace it if it appears dirty or clogged. In dusty environments, you may need to replace it more frequently.

Spark plugs play a vital role in your 442's ignition system, and their condition can significantly impact engine performance. For most 442 models, spark plugs should be replaced every 30,000 miles or when signs of wear appear. Signs of worn spark plugs include rough idling, difficulty starting, and decreased fuel efficiency. When replacing spark plugs, always use the correct type specified for your 442 model and adjust the gap according to the manufacturer's specifications.

The carburetor is another critical component that requires regular attention. A properly tuned carburetor can improve fuel efficiency by up to 15% and increase horsepower, letting your 442 perform at its best. Cleaning the carburetor every 15,000 miles can prevent build-up that affects its performance. If you're not comfortable with carburetor tuning, it's worth having a professional mechanic familiar with classic muscle cars handle this task.

Lastly, don't overlook your 442's cooling system. An overheating engine can cause severe damage, potentially leading to costly repairs. Flush the cooling system every two years to prevent corrosion and extend the life of your 442's engine. This involves draining the old coolant, flushing the system with a cleaning solution, and refilling with fresh coolant. Also, regularly inspect hoses and belts for signs of wear or cracking, replacing them as needed.

Remember, preventive maintenance is always less expensive than major repairs. By following these engine maintenance practices, you'll keep your Oldsmobile 442's powerplant running smoothly, ensuring it continues to deliver the thrilling performance that made it a muscle car icon. Your dedication to proper engine care will not only maintain the value of your 442 but also ensure many more years of exhilarating drives.

Section 10.2: Transmission and Drivetrain Care

The transmission and drivetrain are crucial components of your Oldsmobile 442, responsible for transferring the engine's power to the wheels. Proper maintenance of these systems is essential for optimal performance and longevity of your classic muscle car.

Transmission fluid changes and checks are paramount to the health of your 442's transmission. For most 442 models equipped with the Turbo 400 transmission, it's recommended to change the fluid every 30,000 miles or when it becomes dark or gritty.

Regular checks can help you identify potential issues before they become significant problems. When changing the fluid, be sure to use the correct type specified for your model year. For example, a 1970 442 with a Turbo 400 transmission typically requires Dexron II or III automatic transmission fluid.

For those lucky enough to own a 442 with a manual transmission, clutch maintenance is crucial. A well-maintained clutch can last up to 100,000 miles, but driving habits can significantly affect its lifespan. Signs of a worn clutch include slipping, difficulty shifting, or a high engagement point. When it's time for replacement, consider upgrading to a performance clutch if you frequently engage in high-performance driving.

Differential fluid changes are often overlooked but are vital for the longevity of your 442's rear end. Changing the differential fluid every 30,000 to 50,000 miles can prevent wear and extend the life of this crucial component. When changing the fluid, inspect the differential for any signs of metal shavings, which could indicate internal wear.

Regular drive shaft and U-joint inspections are essential for smooth power delivery and preventing vibrations. Greasing U-joints during every oil change can prevent premature wear and costly repairs. When inspecting, look for any signs of play in the U-joints or unusual wear patterns on the drive shaft.

Axle seal replacement is another critical maintenance task. Leaking axle seals can lead to differential damage if not addressed promptly. Signs of a leaking axle seal include oil spots under the car near the wheels or oil on the inside of the tire. If you notice these symptoms, it's crucial to replace the seals as soon as possible to prevent contamination of the brake system and potential differential failure.

When performing any transmission or drivetrain maintenance, it's essential to use high-quality parts and fluids. While it may be tempting to cut costs, using inferior components can lead to premature wear and reduced performance. Remember, your 442 was built with performance in mind, and maintaining that performance requires quality care.

For those less experienced with mechanical work, it's advisable to seek the help of a professional mechanic who specializes in classic cars. They will have the knowledge and tools necessary to maintain your 442's transmission and drivetrain properly.

Regular maintenance of your transmission and drivetrain not only ensures optimal performance but also helps maintain the value of your classic 442. A well-maintained drivetrain can make a significant difference when it comes time to sell or show your car.

In conclusion, proper care of your 442's transmission and drivetrain is essential for maintaining its performance, reliability, and value. By following these maintenance guidelines and addressing issues promptly, you can ensure that your classic Oldsmobile continues to deliver the exhilarating driving experience it was designed for, mile after mile.

Section 10.3: Suspension and Steering Maintenance

Maintaining the suspension and steering systems of your Oldsmobile 442 is crucial for ensuring a smooth ride, precise handling, and overall safety. These components play a vital role in your car's performance and driver comfort, making their upkeep essential for any 442 enthusiast.

Let's start with shock absorbers, the unsung heroes of your car's suspension system. Shocks are responsible for dampening the bouncing motion of your car's springs, providing a comfortable ride, and maintaining proper tire contact with the road.

Over time, shocks wear out, leading to a bouncy ride, nose-diving during braking, and potential safety issues. As a general rule, it's wise to replace your 442's shocks every 50,000 miles. However, if you notice any signs of wear, such as fluid leaks, excessive bouncing, or uneven tire wear, it's time for a replacement. High-quality shocks can significantly improve your 442's handling and prevent premature tire wear, making them a worthwhile investment.

Next, let's discuss ball joints, which are critical components of your car's suspension and steering systems. These spherical bearings connect the control arms to the steering knuckles, allowing for smooth pivoting motion of the wheels. When ball joints wear out, you may notice symptoms such as clunking noises when driving over bumps, uneven tire wear, or a feeling of looseness in the steering. Regular inspection of ball joints is crucial, as worn ball joints can cause steering wandering and, in extreme cases, complete failure, leading to loss of control. If you notice any play in the ball joints during inspection or hear unusual noises, it's time for a replacement.

Steering linkage lubrication is often overlooked but is vital for maintaining smooth and responsive steering. The steering linkage comprises various components, including tie rod ends, idler arms, and pitman arms. These parts are subject to constant movement and can wear out if not properly lubricated. It's recommended to lubricate these components every 3,000 miles or during each oil change. Using a high-quality grease and ensuring each grease fitting is properly serviced can prevent premature wear and maintain a crisp steering feel.

Alignment is another crucial aspect of suspension maintenance. Proper alignment ensures that your 442's wheels are positioned correctly relative to each other and to the car's body. Misalignment can cause a host of issues, including uneven tire wear, poor fuel economy, and handling problems.

Factors such as hitting potholes, minor accidents, or normal wear and tear can affect your alignment. It's a good practice to have your alignment checked annually, or whenever you notice symptoms such as the car pulling to one side or uneven tire wear. A proper alignment can improve fuel economy by up to 10% and significantly extend the life of your tires, making it a cost-effective maintenance procedure.

Lastly, let's talk about bushings. These often-overlooked components are essential for isolating vibrations and allowing controlled movement between various suspension parts. Over time, bushings can deteriorate, leading to excessive movement, squeaks, and rattles. Worn bushings can cause a multitude of issues, including excessive body roll, poor handling, and uneven tire wear. Inspecting your 442's bushings regularly and replacing them when signs of wear appear can dramatically improve your car's handling characteristics and ride comfort.

Remember, maintaining your 442's suspension and steering systems is not just about performance; it's about safety, too. These systems are critical for maintaining control of your vehicle, especially during emergency maneuvers or adverse weather conditions. Regular inspections, timely replacements, and proper maintenance will ensure your classic Oldsmobile 442 continues to deliver the thrilling driving experience it was designed for, while keeping you and your passengers safe on the road.

By paying attention to these key areas of suspension and steering maintenance, you'll keep your 442 handling like it did when it first rolled off the assembly line, preserving both its performance capabilities and its value as a classic muscle car.

Section 10.4: Brake System Maintenance

Maintaining your Oldsmobile 442's brake system is crucial for both safety and performance. A well-maintained brake system ensures your classic muscle car stops as impressively as it accelerates. Let's dive into the essential aspects of brake system maintenance for your 442.

Brake pad and shoe replacement is a fundamental part of brake system upkeep. The frequency of replacement depends on your driving habits and the type of pads or shoes installed. High-quality brake pads can last up to 70,000 miles, but aggressive driving or frequent stop-and-go traffic can significantly reduce this lifespan. Pay attention to any squealing noises when braking, as this is often an indicator that your pads are wearing thin. When replacing brake pads or shoes, always opt for quality parts that meet or exceed OEM specifications. This ensures optimal performance and longevity.

Brake fluid is often overlooked, but it plays a crucial role in your 442's stopping power. Over time, brake fluid absorbs moisture from the air, which can lead to corrosion within the brake system and reduced braking efficiency. Flushing your brake fluid every two years is a good rule of thumb. This process involves removing all the old fluid from the system and replacing it with fresh, clean fluid. Regular brake fluid changes will maintain a consistent pedal feel and prevent internal corrosion of brake components.

Inspecting your brake lines is another critical maintenance task. Brake lines are susceptible to corrosion, especially in areas where road salt is used during the winter months. A visual inspection of your brake lines should be performed at least once a year. Look for any signs of rust, cracks, or fluid leaks. If you notice any of these issues, replace the affected brake lines immediately. Corrosion or damage to brake lines can lead to sudden brake failure, a potentially extremely dangerous situation. When replacing brake lines, consider upgrading to stainless steel lines for improved durability and performance.

For 442 models equipped with drum brakes, proper adjustment is key to maintaining optimal braking performance. Drum brakes have a self-adjusting mechanism, but over time, this can become less effective. Manual adjustment may be necessary to ensure the brake shoes make proper contact with the drum surface. A properly adjusted drum brake will provide better stopping power and more even wear on the brake shoes. If you're not comfortable performing this adjustment yourself, have it done by a professional mechanic familiar with classic car brakes.

Brake rotors are another component that requires attention. Over time, brake rotors can become warped or develop uneven wear patterns, leading to pulsation in the brake pedal and reduced braking efficiency. Resurfacing the rotors can restore smooth braking performance. However, there's a limit to how many times rotors can be resurfaced. Each rotor has a minimum thickness specification, and once this is reached, replacement is necessary. When replacing rotors, consider upgrading to high-performance cross-drilled or slotted rotors for improved heat dissipation and braking performance.

Regular inspection of all brake components is crucial. This includes checking the brake calipers for smooth operation and any signs of sticking or uneven wear. The brake master cylinder should also be inspected for any leaks or signs of failure. A spongy brake pedal or a gradually sinking pedal can indicate issues with the master cylinder.

Lastly, don't forget about the parking brake. The parking brake mechanism can seize up if not used regularly, especially in areas with high humidity or salt exposure. Make it a habit to use your parking brake regularly to keep the mechanism functioning correctly.

Maintaining your 442's brake system is not just about safety; it's also about preserving the car's originality and value. However, when it comes to brakes, performance and safety should always take precedence over strict originality. If upgrading to modern brake components will significantly improve your 442's stopping power and safety, it's a modification worth considering.

Remember, a well-maintained brake system is essential for enjoying your Oldsmobile 442 to its fullest potential. Regular inspections, timely replacements, and proper adjustments will ensure your classic muscle car stops as impressively as it goes, providing you with confidence and peace of mind every time you hit the road.

Section 10.5: Electrical System Maintenance

Maintaining the electrical system of your Oldsmobile 442 is crucial for ensuring reliable operation and preventing unexpected breakdowns. A well-maintained electrical system not only keeps your car running smoothly but also preserves its originality and value.

Let's start with battery maintenance and replacement. Your 442's battery is the heart of its electrical system, providing the power needed to start the engine and run various components. Regularly inspecting the battery is essential. Check for any signs of corrosion on the terminals and clean them with a mixture of baking soda and water if necessary. Ensure the battery is securely mounted to prevent vibration damage.

In most cases, a well-maintained battery can last up to five years, but extreme temperatures can significantly shorten its lifespan. If you notice slow cranking, dimming lights, or electrical issues, it may be time for a replacement. When selecting a new battery, make sure it matches the original specifications for your 442 model year.

Next, let's discuss alternator testing and replacement. The alternator is responsible for charging the battery and powering the electrical system while the engine is running. A failing alternator can cause a range of issues, from dimming lights to complete electrical failure. To test your alternator, use a multimeter to check the voltage at the battery with the engine running. It should read between 13.8 and 14.2 volts. If the reading is outside this range, your alternator may need attention. Signs of a failing alternator include a squealing noise from the engine bay, flickering lights, or a battery warning light on the dashboard. If you suspect alternator issues, have it professionally tested and replaced if necessary.

The starter motor is another critical component of your 442's electrical system. It's responsible for cranking the engine when you turn the key. Over time, starter motors can wear out, resulting in slow cranking or a clicking sound when attempting to start the car. Regular maintenance of the starter motor includes cleaning the connections and ensuring they're tight. If you hear grinding noises when starting the vehicle, it could indicate worn starter gears, necessitating replacement. To extend the life of your starter, avoid cranking the engine for extended periods if it doesn't start immediately.

Wiring harness inspection and repair is often overlooked, but is crucial for the longevity of your 442's electrical system. Over time, wiring can become brittle, cracked, or damaged, leading to shorts or open circuits. Regularly inspect visible wiring for signs of wear, paying special attention to areas near heat sources or where wires may rub against metal surfaces. Suppose you notice any damaged wiring, repair or replace it immediately. Damaged wiring can cause intermittent electrical issues and, in worst-case scenarios, even lead to fires. Consider using period-correct cloth-covered wiring for replacements to maintain authenticity.

Finally, let's address ignition system maintenance. The ignition system in your 442 is responsible for providing the spark that ignites the fuel mixture in the engine. Regular maintenance of this system is crucial for optimal performance. Start by inspecting the distributor cap and rotor for signs of wear or cracking. These components should be replaced every 50,000 miles or when signs of deterioration appear. Replacing them can prevent misfires and improve engine performance. Check the ignition wires for any cracks or damage, and replace them if necessary. Many 442 enthusiasts prefer to use high-quality, performance-oriented ignition components to enhance reliability and power output.

Remember, when working on your 442's electrical system, always disconnect the battery before starting any work to prevent accidental shorts or shocks. If you're not comfortable performing these maintenance tasks yourself, don't hesitate to consult a professional mechanic experienced with classic cars.

By maintaining your 442's electrical system diligently, you'll ensure that your classic muscle car starts reliably, runs smoothly, and retains its value for years to come. A well-functioning electrical system is not just about convenience; it's about preserving the authentic driving experience that makes the Oldsmobile 442 such a cherished piece of automotive history.

Section 10.6: Body and Interior Care

Maintaining the body and interior of your Oldsmobile 442 is just as crucial as keeping the mechanical components in top condition. A well-preserved exterior and interior not only enhance the car's aesthetic appeal but also contribute significantly to its overall value. Let's dive into the essential aspects of body and interior care for your classic muscle car.

Paint maintenance and protection are paramount in preserving your 442's exterior. Regular washing is the foundation of paint care, removing dirt, grime, and contaminants that can damage the finish over time. Use a high-quality car shampoo and soft microfiber cloths to avoid scratching the paint. After washing, apply a good-quality wax every three months to protect the paint from oxidation and environmental damage. For those looking for long-term protection, consider using a ceramic coating, which can provide years of protection with proper maintenance.

Rust prevention and treatment are critical for any classic car, especially in regions with harsh winters or high humidity. Inspect your 442 regularly for any signs of rust, paying close attention to common problem areas such as wheel wells, floor pans, and lower body panels. If you spot any rust, address it immediately to prevent it from spreading. For small spots, you can sand down to bare metal, apply a rust converter, and touch up with matching paint. For more extensive rust issues, professional body shop intervention may be necessary. Applying a rust inhibitor to vulnerable areas can prevent costly body repairs in the future.

Chrome and trim care is essential in maintaining your 442's classic look. These components are susceptible to pitting, oxidation, and discoloration if not correctly maintained. Use a dedicated chrome cleaner to restore shine and prevent pitting. For badly oxidized chrome, you may need to use a metal polish, but be cautious not to over-polish, as this can remove the chrome plating. For anodized aluminum trim, use a gentle cleaner to avoid damaging the finish. Regular cleaning and polishing will keep your 442's brightwork looking its best.

Interior cleaning and preservation are crucial for maintaining both the appearance and value of your 442. Different materials require different care approaches. For leather seats, use a leather cleaner followed by a conditioner every six months to prevent cracking and maintain suppleness. Vinyl surfaces should be cleaned with a vinyl-specific product and protected with a UV-resistant dressing to avoid fading and cracking. Cloth upholstery can be vacuumed regularly and shampooed as needed. For the dashboard and other hard surfaces, use a gentle cleaner and avoid silicone-based products, as they can leave a greasy residue. Don't forget to clean and condition the steering wheel, shift knob, and other frequently touched surfaces.

Weatherstripping replacement is often overlooked but is crucial for both comfort and preservation. Worn or damaged weatherstripping can lead to water leaks, increased wind noise, and even damage to the car's interior. Inspect the weatherstripping around doors, windows, and the trunk on a regular basis. If you notice any cracks, tears, or signs of hardening, it's time for replacement. Using high-quality, model-specific weatherstripping will ensure a proper fit and seal. Properly maintained weatherstripping will keep your 442's interior dry, quiet, and protected from the elements.

Lastly, don't forget about the often-neglected areas like the trunk and the under-hood regions. Keep the trunk clean and dry to prevent rust and maintain the spare tire and jack in good condition. Under the hood, clean the engine bay periodically to make it easier to spot leaks and other issues. Use a degreaser for oily areas and a general-purpose cleaner for different surfaces. After cleaning, apply a protectant to rubber and plastic components to prevent them from drying out and cracking.

By following these body and interior care tips, you'll ensure that your Oldsmobile 442 looks as good as it performs. Remember, a well-maintained classic car is not just a joy to drive; it's a valuable asset that appreciates over time. Your dedication to preserving both the mechanical and aesthetic aspects of your 442 will pay dividends in driving enjoyment and long-term value.

Section 10.7: Preventive Maintenance Schedule

A well-structured preventive maintenance schedule is the key to keeping your Oldsmobile 442 in prime condition. By following a routine, you can catch potential issues early, prevent costly repairs, and ensure your classic muscle car remains road-ready and performing at its best.

Daily and weekly checks form the foundation of your maintenance routine. Each time you drive your 442, take a moment to perform a quick visual inspection. Check for any fluid leaks underneath the car, ensure all lights are functioning correctly, and listen for any unusual noises during operation. Weekly, check tire pressures and inspect the tires for any signs of uneven wear or damage. These simple checks can alert you to developing problems before they become serious.

Monthly maintenance tasks are your next line of defense. Once a month, pop the hood and check all fluid levels, including engine oil, coolant, power steering fluid, and brake fluid. Top up as necessary, but be sure to investigate any significant drops in fluid levels, as they may indicate a leak or other issue. This is also a good time to inspect belts and hoses for signs of wear or cracking. Don't forget to test your battery's charge and clean any corrosion from the terminals.

Quarterly, or every three months, it's time for a more thorough inspection. Check your brake system, including pads, rotors, and brake lines. Lubricate door hinges, hood latches, and other moving parts. If you're comfortable doing so, rotate your tires to ensure even wear. This is also an excellent time to wash and wax your 442, paying close attention to the undercarriage to remove any built-up road grime or salt.

Your annual maintenance checklist should be comprehensive. Consider this your 442's yearly physical. Start with an oil and filter change, even if you haven't hit the mileage threshold; old oil can become acidic and harm your engine. Replace the air filter and spark plugs if needed. Flush and replace your coolant and brake fluid. Check and adjust the timing and idle speed. Inspect and grease all suspension components. Don't forget to change the transmission fluid and differential oil if due.

This is also the time for a thorough inspection of often-overlooked items. Check the condition of your weatherstripping, inspect the exhaust system for leaks or damage, and test all electrical systems, including the alternator and starter. If you have access to a lift, this is an excellent opportunity to inspect the underside of your 442 for any signs of rust or structural issues.

Long-term storage preparation is crucial if you plan to store your 442 for an extended period, such as over winter. Start by giving your car a thorough wash and wax to protect the paint. Fill the gas tank and add a fuel stabilizer to prevent the gasoline from deteriorating. Change the oil to remove any contaminants that could damage your engine during storage. Remove the battery and store it in a warm, dry place, or connect it to a trickle charger.

Slightly overinflate the tires to prevent flat spots, or better yet, put the car on jack stands to take the weight off the tires and suspension. Use a high-quality car cover to protect your 442 from dust and moisture. Start the engine and let it run to operating temperature once a month to keep seals lubricated and prevent moisture buildup.

Remember, preventive maintenance is an investment in your 442's future. By following this schedule and addressing issues promptly, you'll keep your classic Oldsmobile running smoothly, looking great, and appreciated for years to come. Your dedication to maintenance will pay off every time you turn the key and hear that mighty V8 roar to life, ready for another thrilling drive.

High Octane Heritage: *Celebrating the Oldsmobile 442*

High Octane Heritage: *Celebrating the Oldsmobile 442*

Chapter 11: The 442 Community: Stories from Owners and Enthusiasts

Section 11.1: The Birth of Passion - First Encounters with the 442

The Oldsmobile 442 has a unique ability to capture hearts and ignite passions like few other cars. For many enthusiasts, their first encounter with this iconic muscle car was a defining moment, one that would shape their automotive journey for years to come. This section explores the diverse and often serendipitous ways in which 442 lovers first discovered their automotive soulmate.

For some, it was truly love at first sight. Tom Jensen, a longtime 442 owner from Michigan, recalls the moment he first laid eyes on a 1970 442 W-30 at a local dealership. "I was walking past Oldsmobile's showroom window, and there it was, Sebring Yellow with black stripes. I stopped dead in my tracks. The aggressive stance, the bulging hood, the gleaming chrome, it was the most beautiful car I'd ever seen. I knew right then that I had to have one someday."

Many enthusiasts speak of childhood dreams finally realized. Sarah Martinez grew up hearing her father's stories about the 1968 442 he owned in his youth. "Dad would always get this far-off look in his eyes when he talked about that car," she remembers. "I made a promise to myself that I'd buy him one for his 60th birthday. It took years of saving and searching, but seeing his face when I handed him the keys to a restored '68 442 made it all worth it."

Sometimes, the discovery of a 442 comes as a surprise. John Teller shares his story of stumbling upon a hidden gem: "I was helping clean out my great-uncle's old barn after he passed away. Under a tarp in the corner, covered in decades of dust, was this neglected 1965 442. I had no idea he even owned one! It was in rough shape, but I saw the potential. That barn find started my restoration journey."

Family legacies play a significant role in many 442 enthusiasts' stories. Mike Dolan's connection to the 442 spans three generations. "My grandfather bought a new '66 442 and eventually passed it down to my dad. I grew up in the back seat of that car, going to shows and cruises. When Dad handed me the keys on my 21st birthday, it felt like I was inheriting more than just a car; it was our family's history."

The hunt for the perfect 442 can be an adventure in itself. Lisa Chen, a California-based collector, spent years searching for her dream car. "I knew exactly what I wanted, a 1969 Hurst/Olds in Cameo White with gold stripes. I scoured classifieds, attended countless auctions, and networked with other collectors. When I finally found one in original condition, I flew across the country the next day to make the deal. The thrill of the hunt made owning it even sweeter."

These stories of first encounters with the 442 highlight the car's enduring appeal and its ability to form lasting connections with enthusiasts. Whether it was a chance sighting, a family tradition, or a long-awaited dream come true, each narrative underscores the emotional impact of discovering the 442. For many, that initial encounter was just the beginning of a lifelong passion, setting the stage for years of enjoyment, restoration projects, and community involvement within the world of Oldsmobile enthusiasts.

As we'll see in the following sections, these first encounters often led to deeper involvement with the 442, from painstaking restorations to exhilarating drives and the forging of lasting friendships. The birth of passion for the 442 is not just about appreciating a classic muscle car; it's about becoming part of a rich automotive legacy that continues to thrive today.

Section 11.2: Restoration Journeys - Bringing 442s Back to Life

The journey of restoring an Oldsmobile 442 is often as thrilling as driving one. For many enthusiasts, the process of bringing these classic muscle cars back to their former glory is a labor of love that creates lasting memories and a deep sense of accomplishment. This section examines the diverse experiences of 442 owners who have undertaken restoration projects, ranging from miraculous barn finds to multi-generational family efforts.

One of the most exciting stories in the 442 community revolves around barn find miracles. Take, for instance, the tale of Mark Thompson, a devoted 442 fan from Ohio. In 2015, Mark stumbled upon a tip about an abandoned 1970 442 W-30 hidden away in an old barn just outside of Dayton.

Upon investigation, he discovered a dust-covered beauty that had been sitting untouched for over three decades. "It was like finding buried treasure," Mark recalls. "The car was in surprisingly good condition, considering its long slumber. The original Viking Blue paint was still visible under layers of dust, and the numbers-matching engine was intact." Mark spent the next two years meticulously restoring the car, bringing it back to its original showroom condition.

Restoration projects often become powerful bonding experiences, especially between fathers and sons. The story of the Johnsons from Texas is a prime example. Jim Johnson had always dreamed of owning a 1968 442, the same model year he admired in his youth.

When he finally found one in need of extensive restoration, he saw it as an opportunity to connect with his teenage son, Mike. "At first, Mike was more interested in modern cars," Jim explains. "But as we started working on the 442 together, he began to appreciate the craftsmanship and engineering of these classic machines."

Over three years, the Johnsons spent countless weekends in their garage, learning together as they tackled everything from engine rebuilding to interior refurbishment. The project not only resulted in a beautifully restored 442 but also strengthened their relationship and ignited a passion for classic cars in the younger Johnson.

For many 442 enthusiasts, the restoration process is a steep learning curve, especially for first-time restorers. Sarah Martinez, a software engineer from California, shares her experience restoring a 1969 442 convertible as her first major automotive project. "I had basic mechanical knowledge, but restoring a classic car was a whole new ballgame," Sarah admits. "I made plenty of mistakes along the way, like ordering the wrong parts or underestimating the time certain tasks would take."

However, Sarah persevered, relying on online forums, restoration manuals, and the guidance of experienced 442 owners in her local car club. After four years of hard work, her perseverance paid off with a stunning Trumpet Gold 442 that she now proudly drives to car shows across the state.

Professional restorations also play a significant role in preserving 442 history. Renowned restorer Bob Anderson of Classic Muscle Restorations in Michigan has worked on dozens of 442s over his 30-year career. He shares insights on one of his most notable projects, a rare 1970 442 W-30 with the even rarer W-27 aluminum differential. "This car was a challenge because of its rarity and the owner's insistence on absolute originality," Bob explains.

"We had to source NOS (New Old Stock) parts from across the country and even had to recreate some components that were no longer available." The restoration took 18 months and over 2,000 hours of labor, but the result was a concours-quality 442 that went on to win multiple awards at prestigious car shows.

The dramatic transformations achieved through restoration never fail to impress. Tom and Linda Ramirez from Florida showcase the before and after of their 1966 442, which they found rusting away in a junkyard. "When we first saw it, most people would have considered it beyond saving," Linda recalls.

"The floor pans were rusted through, the engine was seized, and the interior was home to a family of raccoons." Undeterred, the couple spent five years bringing the car back to life. Today, their Autumn Bronze 442 is a showstopper, with its gleaming paint, immaculate interior, and perfectly detailed engine bay. "Seeing the car now, it's hard to believe it's the same one we pulled out of that junkyard," Tom says proudly.

These restoration journeys highlight the dedication, skill, and passion that 442 owners bring to their projects. Whether it's a barn find miracle, a family bonding experience, a learning adventure for a first-time restorer, or a professional undertaking, each restoration adds another chapter to the rich history of the Oldsmobile 442. These projects do more than just preserve classic cars; they keep the spirit of American muscle alive and create a new generation of enthusiasts committed to carrying on the 442 legacy.

Section 11.3 The Thrill of the Drive - On the Road with the 442

There's nothing quite like the experience of driving an Oldsmobile 442, and the owners we spoke to were eager to share their exhilarating stories from behind the wheel. From daily commutes to cross-country adventures, these muscle car enthusiasts have put their 442s through their paces, creating lasting memories along the way.

For some, the 442 isn't just a weekend cruiser; it's a daily driver. John Thompson, a software engineer from California, has been using his 1970 442 as his primary vehicle for over a decade. "There's something special about starting each day with that rumble," John says. "Sure, it's not the most practical choice, but it puts a smile on my face every single morning. The looks I get from other commuters are priceless."

Others have taken their love for the 442 to the extreme, embarking on cross-country road trips that test both car and driver. Sarah and Mike Rodriguez, a retired couple from Florida, spent six weeks driving their restored 1968 442 convertible along Route 66.

"We've always dreamed of doing the Mother Road, and we couldn't think of a better car to do it in," Sarah explains. "The 442 performed flawlessly, and we made so many new friends along the way. Other classic car owners would spot us at gas stations and strike up conversations. It was like being part of a rolling car show."

For the more competitive 442 owners, track days provide the ultimate thrill. Tom Lawson, a mechanic from Michigan, frequently takes his modified 1969 442 to local drag strips. "There's nothing like launching a 442 down the quarter-mile," Tom grins. "I've put a lot of work into this car, and seeing it perform on the track is incredibly satisfying. It's not just about the speed - it's about carrying on the legacy of what these cars were built for."

Car shows are another popular venue for 442 owners to showcase their prized possessions. Linda Chen, a restoration specialist from Texas, has won numerous awards with her immaculately preserved 1965 442. "The first time I won 'Best in Show,' I was on cloud nine," Linda recalls. "It validated all the hard work and attention to detail I'd put into the restoration. But more than that, it was a chance to educate younger generations about these amazing machines."

Of course, not all memorable moments behind the wheel of a 442 are planned. Many owners shared stories of unexpected adventures and chance encounters. Mark Sullivan, a teacher from New York, still chuckles when he remembers the day his 1971 442 caught the eye of a Hollywood location scout. "I was just filling up at a gas station when this guy runs over, asking if I'd be interested in having my car featured in a period film they were shooting," Mark says. "Next thing I know, I'm watching my 442 share the screen with some A-list actors. It was surreal."

These stories highlight the diverse experiences of 442 owners, but they all share a common thread: the unbridled joy of driving these iconic muscle cars. Whether it's the daily commute, a cross-country odyssey, or a heart-pounding run down the drag strip, the 442 continues to deliver thrills and create lasting memories for its devoted owners.

The passion these enthusiasts have for their 442s is palpable, and it's clear that for many, the car is much more than just a mode of transportation. It's a time machine, a conversation starter, a source of pride, and a constant adventure. As Tom Lawson aptly put it, "Every time I turn the key, I'm not just starting an engine - I'm continuing a legacy."

Section 11.4: The 442 Community - Bonds Forged Through Passion

The Oldsmobile 442 is more than just a car; it's the heart of a vibrant and passionate community. Across the country and around the world, 442 enthusiasts have formed tight-knit groups, bonded by their shared love for this iconic muscle car. These communities serve as hubs of knowledge, support, and friendship, playing a crucial role in preserving the legacy of the 442.

Local 442 clubs have become the backbone of the community, providing enthusiasts with an opportunity to connect with like-minded individuals in their area. The Motor City Rockets, based in Detroit, is one such group that has been active for over three decades. Club president Mike Thompson explains, "Our monthly meetups are like family reunions. We share stories, swap parts, and help each other with restoration projects. It's not just about the cars; it's about the people."

In recent years, the internet has revolutionized the way 442 enthusiasts connect. Online forums and social media groups have created virtual communities that transcend geographical boundaries. The "442 Nation" Facebook group, with over 20,000 members, has become a go-to resource for owners seeking advice, sharing photos, and discussing all things 442. As member Sarah Johnson puts it, "I can post a question about a carburetor issue at 2 AM, and within minutes, I'll have multiple responses from experts around the world. It's incredible."

Annual meetups and conventions serve as the pinnacle of 442 community gatherings. The Oldsmobile Nationals, held in different locations each year, draws hundreds of 442 owners and thousands of spectators. These events feature car shows, swap meets, and technical seminars, but more importantly, they provide a chance for online friends to meet in person and forge lasting relationships. Tom Garcia, a regular attendee, shares, "I've made lifelong friends at these conventions. We stay in touch year-round and plan road trips to visit each other's garages."

One of the most beautiful aspects of the 442 community is the spirit of mentorship and knowledge sharing. Experienced owners often take newcomers under their wing, guiding them through the intricacies of 442 ownership and restoration. Veteran restorer Bob

Williams recalls, "I remember being a clueless kid with a rusted-out 442. Now, I make it a point to help young enthusiasts avoid the mistakes I made. Seeing their faces light up when they fire up their restored 442 for the first time, that's what it's all about."

The sense of camaraderie within the 442 community often leads to collaborative projects. In 2019, a group of enthusiasts from across the Midwest came together to restore a rare 1970 W-30 442 for a terminally ill fellow member. The project, dubbed "Operation W-30," saw dozens of volunteers donating time, parts, and expertise to complete the restoration in record time. Project coordinator Lisa Martinez reflects, "It was a labor of love. We worked nights and weekends to make sure our friend could enjoy his dream car. That's the power of this community."

The 442 community also plays a vital role in preserving the car's history and educating the public. Many clubs organize displays at local car shows and participate in community events to showcase the 442's legacy. The Southern California 442 Club, for instance, hosts an annual "442 Day" at a local high school, where members bring their cars and talk to students about automotive history and restoration techniques.

As the years pass and the 442 becomes increasingly rare, the importance of this passionate community grows. It's not just about maintaining cars; it's about keeping alive the spirit of an era. The 442 community ensures that the knowledge, stories, and passion associated with these machines are passed down to future generatlons.

In the words of longtime 442 owner and community leader Frank Thompson, "The 442 brought us together, but it's the friendships and shared experiences that keep us coming back. This community is a testament to the enduring appeal of the 442 and the bond it creates among enthusiasts. As long as there are 442s on the road, there will be a community of passionate individuals ready to support, celebrate, and preserve this incredible piece of automotive history.

Section 11.5: Preserving History - Collectors and Their Prized 442s

The world of Oldsmobile 442 collectors is a fascinating realm where passion meets preservation. These dedicated individuals play a crucial role in maintaining the legacy of this iconic muscle car, often going to great lengths to acquire, restore, and showcase their prized possessions.

One such collector is Tom Hanson, proud owner of a rare 1970 Oldsmobile 442 W-30 convertible. "When I found this car, I knew I had stumbled upon something special," Tom recalls. "Only 264 W-30 convertibles were produced that year, and finding one in restorable condition was like hitting the jackpot." Tom's story is not uncommon among 442 collectors, who often spend years searching for that elusive, perfect example to add to their collection.

The quest for originality is a driving force for many collectors. Sarah Martinez, who maintains a collection of five 442s from various model years, emphasizes the importance of authenticity. "It's not just about having a beautiful car," she explains. "It's about preserving a piece of automotive history exactly as it left the factory." Sarah's dedication to originality extends to using period-correct parts and materials in her restorations, even if it means spending months tracking down a specific component.

For some collectors, the 442 represents more than just a hobby; it's a serious investment. Michael Chen, a financial advisor and 442 enthusiast, offers insight into the investment potential of these classic muscle cars. "Over the past decade, we've seen a steady increase in the value of well-maintained 442s, particularly the rarer models," Michael notes. "While I collect primarily out of passion, it's reassuring to know that these cars can also be a sound financial investment if properly cared for."

The preservation of 442 history often extends beyond individual collections to more formal settings. Robert Thompson, curator of the American Muscle Car Museum in Florida, showcases several significant 442 models in his facility. "We have everything from an original 1964 4-4-2 to a 1972 Hurst/Olds," Robert proudly states. "Our goal is to educate visitors about the 442's place in muscle car history and its impact on American car culture."

As these collectors age, many are giving serious thought to the future of their beloved 442s. John Williamson, a longtime collector in his seventies, shares his perspective: "I've spent most of my life building this collection, and now my focus is on ensuring these cars will be appreciated and cared for after I'm gone." John has been working with his children and grandchildren, instilling in them the same passion he has for these classic Oldsmobiles.

The world of 442 collecting is not without its challenges. The increasing rarity of original parts and the rising costs of restoration can be daunting. However, for these dedicated enthusiasts, the rewards far outweigh the difficulties. "Every time I slide behind the wheel of my '68 442," says collector Lisa Patel, "I'm transported back in time. It's like being a custodian of history, and that feeling is priceless."

From rare finds to museum-quality collections, from the quest for originality to investment considerations, 442 collectors are united in their commitment to preserving these remarkable machines. Their efforts ensure that future generations will have the opportunity to experience the power, style, and historical significance of the Oldsmobile 442. As these collectors pass the torch to the next generation of enthusiasts, they do so with the knowledge that they've played a vital role in keeping the 442 legacy alive and thriving.

High Octane Heritage: *Celebrating the Oldsmobile 442*

Section 11.6: Racing Legends - 442 Owners in Motorsports

The Oldsmobile 442's reputation as a muscle car icon wasn't just built on the streets—it was also forged on the race track. In this section, we delve into the thrilling world of 442 owners who took their passion for speed and competition to the next level.

Drag racing has always been a natural fit for the 442's powerful engine and straight-line acceleration. Many owners have fond memories of racing their 442s at local drag strips. Tom Johnson, a longtime 442 enthusiast from California, recalls, "My '68 442 was my weekend warrior. I'd drive it to work all week, then hit the strip on Saturday nights. The roar of that 400 cubic inch V8 launching off the line was pure adrenaline."

Amateur circuit racing provided another avenue for 442 owners to showcase their cars' performance. Weekend warriors like Sarah Martinez found a perfect balance between daily driving and occasional track time. "I bought my '70 442 as a fun daily driver, but soon found myself entering local time attack events. It wasn't the most nimble car on the track, but it sure was a blast to drive at the limit," Sarah shares.

As 442s have aged into classic status, many owners have discovered the joys of vintage racing. These events allow enthusiasts to experience their cars as they were meant to be driven, in a competitive yet respectful environment. Mark Thompson, who races his '69 442 in vintage Trans-Am events, explains, "There's something special about lining up on the grid with other muscle cars from the era. It's like stepping back in time, but with modern safety equipment."

Some 442 owners have pushed their cars to achieve remarkable feats. In 2005, Bob Anderson's heavily modified 1970 442 broke the 200 mph barrier at the Bonneville Salt Flats, showcasing the 442 platform's potential when taken to its extreme. "It took years of development and a lot of trial and error," Bob recalls. "But crossing that 200 mph mark in a car that started life as a street 442 was an incredible feeling."

The transition from streetcar to dedicated racecar is a journey many 442 owners have undertaken. This process often involves extensive modifications, from engine builds to suspension upgrades and safety enhancements. John Reeves, who campaigned a '69 442 in SCCA competition, describes the evolution: "My 442 started as a stock restoration project, but the more I raced it, the more I modified it. By the end, it was a purpose-built race car that just happened to have 442 bodywork."

These racing endeavors have not only provided thrills for the owners but have also contributed to the 442's enduring legacy. Each race, whether at a local drag strip or a national event, has added to the model's mystique and appeal. The sight and sound of a 442 at full throttle continue to inspire new generations of enthusiasts, ensuring that the model's racing heritage lives on.

As we've seen through these stories, the Oldsmobile 442's racing pedigree is as diverse as its owner base. From grassroots drag racing to high-speed land speed records, 442 owners have continually pushed the boundaries of what these muscle cars can achieve. Their passion for competition has not only provided personal satisfaction but has also played a crucial role in cementing the 442's status as a true performance legend.

Section 11.7: The Next Generation - Young Enthusiasts and the 442's Future

As the Oldsmobile 442 approaches its sixth decade since its introduction, a new generation of enthusiasts is emerging, breathing fresh life into this classic muscle car. These young owners and admirers are not only preserving the 442's legacy but also reimagining its place in modern car culture.

High Octane Heritage: *Celebrating the Oldsmobile 442*

Twenty-eight-year-old Matt Simmons, a software engineer from Austin, Texas, represents this new wave of 442 enthusiasts. "I grew up watching my dad work on his '68 Cutlass," Matt explains. "When I finally got the chance to own a 442, it felt like coming full circle. There's something about the raw power and classic styling that you just can't find in modern cars." Matt's story is increasingly common among younger enthusiasts who are drawn to the 442's timeless appeal and hands-on ownership experience.

While many young owners appreciate the 442's original charm, others are finding ways to blend classic aesthetics with modern technology. Sarah Chen, a 31-year-old automotive designer from Los Angeles, has given her 1970 442 a contemporary twist. "I've installed a digital instrument cluster and a modern infotainment system," she says. "It maintains the classic look, but with conveniences that make it more enjoyable as a daily driver." This trend of tasteful modernization is helping to keep these vintage machines relevant and accessible to a tech-savvy generation.

Social media has played a crucial role in introducing the 442 to new audiences. Instagram influencer Jake Martinez, known as @442_Jake to his 100,000 followers, shares his passion for cars through stunning photography and video content. "I started posting about my restoration project, and it just took off," Jake recalls. "Now, I'm connecting with 442 fans from all over the world. It's amazing to see how many young people are getting excited about these cars."

Recognizing the importance of educating younger generations about automotive history, many 442 clubs and organizations are stepping up their outreach efforts. The Oldsmobile 442 Heritage Foundation, for instance, has launched a program that brings restored 442s to high schools and colleges for hands-on learning experiences. "We want to show young people that there's more to cars than just getting from A to B," says Lisa Tompkins, the Foundation's president. "These machines are a part of our cultural heritage."

As they look to the future, young 442 enthusiasts are optimistic about the car's enduring appeal. "I think the 442 will always have a place in car culture," says 25-year-old mechanic and 442 owner Alex Rodriguez. "It represents an era of American ingenuity and performance that's hard to replicate. As long as there are people who appreciate that, the 442 will live on."

The passion of these young enthusiasts is not just about preserving a piece of automotive history; it's about carrying forward a legacy. They're finding new ways to celebrate the 442, from organizing meets that combine classic car showcases with modern street food and music festivals, to creating YouTube channels dedicated to DIY restoration tips for muscle cars.

Perhaps most encouragingly, many of these young owners see their role as temporary custodians of these machines. "I don't just own this car," says 29-year-old teacher and 442 enthusiast Emily Watson. "I'm preserving it for the next generation. Someday, I hope to pass this car on to someone who will love it as much as I do."

As we look to the future, it's clear that the Oldsmobile 442's legacy is in good hands. These young enthusiasts are not just maintaining these classic muscle cars; they're ensuring that the spirit of the 442, its power, style, and the community it fosters, will continue to inspire for generations to come. In their hands, the 442 isn't just a relic of the past, but a living, evolving legend that bridges the gap between automotive history and the future of car culture.

High Octane Heritage: *Celebrating the Oldsmobile 442*

High Octane Heritage: *Celebrating the Oldsmobile 442*

Chapter 12: Legacy and Comparison: The 442 vs. Other Muscle Car Legends

Section 12.1: The Muscle Car Landscape

The 1960s and early 1970s marked a golden age in American automotive history, known as the muscle car era. This period was characterized by powerful, high-performance vehicles that captured the imagination of car enthusiasts and casual drivers alike. To truly appreciate the Oldsmobile 442's place in this pantheon of automotive legends, we must first understand the landscape in which it thrived.

The muscle car era officially began in 1964 with the introduction of the Pontiac GTO, which set the stage for a new category of vehicles that prioritized performance and speed. This era was defined by mid-size cars equipped with large-displacement V8 engines, creating a perfect balance of power and practicality. The key players in this market included General Motors, Ford, and Chrysler, each vying for supremacy on the streets and drag strips of America.

What truly defined a muscle car went beyond mere horsepower figures. These vehicles were characterized by their aggressive styling, often featuring hood scoops, bold graphics, and wide tires. They were designed to appeal to young, performance-minded buyers who wanted a car that could dominate on the weekends but still serve as a daily driver. The muscle car represented a uniquely American ideal of freedom, power, and individuality.

In this competitive landscape, the Oldsmobile 442 carved out a unique position for itself. While many muscle cars prioritized raw power at the expense of refinement, the 442 struck a balance between performance and luxury. It offered the punch that muscle car enthusiasts craved, but with a level of sophistication that set it apart from its more brutish competitors. This positioning allowed Oldsmobile to attract buyers who wanted the thrill of a muscle car without sacrificing comfort and style.

Consumer expectations during the muscle car boom were high and ever-evolving. Buyers demanded more power, better handling, and increasingly distinctive styling with each passing year. They wanted cars that could win at the drag strip on Saturday night and turn heads at the drive-in on Sunday afternoon. The 442, with its combination of performance, handling, and Oldsmobile's reputation for quality, met these expectations in a way that few other cars could match.

The muscle car landscape was also shaped by a spirit of one-upmanship among manufacturers. Each year brought new models, more powerful engines, and increasingly outrageous styling cues as companies sought to outdo each other and capture the public's attention. This competition drove innovation and pushed the boundaries of what was possible in a production car.

However, this era was not without its challenges. As the 1960s gave way to the 1970s, changing social attitudes, stricter emissions regulations, and rising insurance costs began to put pressure on the muscle car market. Manufacturers had to adapt, finding ways to maintain performance while meeting new standards for safety and fuel efficiency.

Throughout these changes, the Oldsmobile 442 remained a steadfast presence, evolving with the times but never losing sight of its performance heritage. Its ability to adapt while maintaining its core identity was a testament to the model's strength and the vision of its creators.

Understanding this broader context is crucial for appreciating the significance of the 442. It wasn't just another muscle car; it was a sophisticated performer that helped define an era, setting standards for performance, style, and engineering that continue to influence automotive design to this day. As we delve deeper into comparisons with its contemporaries, keep in mind the unique space that the 442 occupied in this dynamic and competitive landscape.

Section 12.2: 442 vs. Pontiac GTO

The rivalry between the Oldsmobile 442 and the Pontiac GTO is a legend in the world of muscle cars. Both vehicles emerged from the General Motors stable, each vying for supremacy in the burgeoning muscle car market of the 1960s. The GTO, often credited with kick-starting the muscle car era, hit the streets in 1964, a year before the 442 became a standalone model. This head start gave Pontiac an early advantage, but Oldsmobile was quick to respond with a formidable contender.

In terms of origins and development, the GTO began as an option package for the Pontiac Tempest, while the 442 started as a performance option for the F-85 and Cutlass models. Both cars evolved from option packages to become distinct models in their own right, showcasing the escalating horsepower wars of the era.

When it comes to engine and performance comparisons, both the 442 and GTO offered impressive powerplants. The GTO initially went with a 389 cubic inch V8, while the 442 boasted a 400 cubic inch engine. As the years progressed, both models saw increases in displacement and power output. The 1970 442 with the W-30 package, featuring a 455 cubic inch V8, was a true powerhouse, producing 370 horsepower and a staggering 500 lb-ft of torque. The GTO countered with its Ram Air IV 400 cubic inch engine, rated at 370 horsepower. In real-world performance, the two were often neck and neck, with the outcome of drag races depending more on driver skill than any significant advantage in raw power.

Styling and design are where these two muscle cars truly diverged. The GTO embraced a more aggressive, youth-oriented look with its split grille, stacked headlights, and muscular stance. The 442, in keeping with Oldsmobile's more mature image, opted for a sleeker, more refined appearance. Its long hood, sculpted body lines, and subtle performance cues appealed to a slightly older demographic while still turning heads at stoplights and drive-ins alike.

In terms of sales figures and market reception, the GTO initially had the upper hand. Its early entry into the market and aggressive marketing campaign, including the famous "GTO Tiger" ads, helped it achieve higher sales numbers in the mid-1960s. However, the 442 steadily gained ground, especially as it refined its performance credentials and luxury appointments. By the late 1960s and early 1970s, the 442 was giving the GTO a run for its money in both performance and sales.

The long-term impact and collector value of both cars have proven to be substantial. The GTO, with its reputation as the original muscle car, has long been a favorite among collectors. However, scarce models like the W-30 package cars, with only 442 produced, have seen their value steadily increase over the years. Today, both vehicles command respect and high prices in the collector car market, with remarkably well-preserved or rare examples fetching six-figure sums at auction.

Ultimately, the 442 vs. GTO rivalry exemplifies the intense competition and rapid evolution of the muscle car era. While the GTO may have fired the first shot in the muscle car wars, the 442 proved to be a worthy adversary, often surpassing its Pontiac cousin in terms of refinement and all-around performance. Both cars left an indelible mark on automotive history, inspiring passion and devotion that continues to this day among muscle car enthusiasts.

Section 12.3: 442 vs. Chevrolet Chevelle SS

The Oldsmobile 442 and the Chevrolet Chevelle SS were two of General Motors' most formidable muscle cars, sharing a common lineage but each carving out its own distinct identity. Both vehicles were built on GM's A-body platform, which provided a solid foundation for high-performance machines. However, the similarities between these two powerhouses only serve to highlight their unique characteristics and the different approaches taken by Oldsmobile and Chevrolet.

At the heart of both cars lay their impressive powerplants. The 442, particularly in its later years, offered the mighty 455 cubic inch V8, while the Chevelle SS countered with its own 454 cubic inch big-block. Both engines were capable of producing prodigious amounts of power, with various states of tune offering anywhere from 360 to over 450 horsepower.

On the drag strip, these cars were remarkably close in performance, with quarter-mile times often separated by mere tenths of a second. The 442, however, had a slight edge in terms of torque, which translated to explosive off-the-line acceleration.

While performance was paramount, the interiors of these muscle cars revealed their different brand philosophies. The 442, true to Oldsmobile's position as a more upscale division of GM, offered a level of refinement and comfort that was a notch above the Chevelle.

Plush seating, wood-grain accents, and a more comprehensive array of gauges were hallmarks of the 442's cabin. The Chevelle SS, while still well-appointed, focused more on a sporty, no-frills approach that appealed to hardcore performance enthusiasts.

Brand perception played a significant role in how these cars were marketed and received. Oldsmobile, with its long-standing reputation for innovation and engineering excellence, positioned the 442 as a sophisticated performance machine for the discerning buyer. The "442" designation itself spoke to the car's technical prowess, originally standing for four-barrel carburetor, four-speed manual transmission, and dual exhausts. Chevrolet, on the other hand, leveraged its broader appeal and racing heritage to present the Chevelle SS as the everyman's muscle car, accessible yet potent.

When it comes to rarity and collectibility, both the 442 and Chevelle SS are highly sought after by enthusiasts and collectors. However, certain factors give the 442 a slight edge in this arena. The Oldsmobile's lower production numbers, particularly for special editions like the W-30 package, have made pristine examples increasingly valuable.

The Chevelle SS, while also useful, was produced in greater numbers, making it more common on the collector market. That said, rarity alone doesn't tell the whole story; the Chevelle's iconic status in popular culture has ensured its place as one of the most recognizable and desirable muscle cars of all time.

Ultimately, the comparison between the Oldsmobile 442 and the Chevrolet Chevelle SS reveals more than just differences in badging and sheet metal. It showcases GM's strategy of offering distinct flavors of performance to cater to different market segments.

The 442 represented the thinking man's muscle car, blending luxury with brute force, while the Chevelle SS embodied the pure, unadulterated essence of the American muscle car ethos. Both have earned their place in the pantheon of automotive legends, each a testament to the golden age of American performance.

Section 12.4: 442 vs. Ford Mustang

The Oldsmobile 442 and the Ford Mustang represent two distinct approaches to American performance cars of the 1960s and 1970s. While the Mustang pioneered the pony car segment, the 442 stood as Oldsmobile's response to the growing demand for high-performance vehicles.

The Mustang's introduction in 1964 sent shockwaves through the automotive industry, creating a new category of compact, affordable sports cars. Oldsmobile, recognizing the threat, positioned the 442 as a more mature, refined alternative to Ford's youthful offering. The 442 package, initially an option for the Cutlass, was Oldsmobile's way of saying, "We can perform too, but with a touch of sophistication."

When it came to engine options and performance metrics, both cars offered a range of powerplants to suit different buyer preferences. The Mustang's engine lineup ranged from economical inline-sixes to powerful V8s, including the legendary 428 Cobra Jet. The 442, true to its numerical namesake, offered a 400 cubic-inch V8 as standard, with options like the mighty 455 cubic-inch V8 in later years. In terms of raw power, the top-tier versions of both cars were closely matched, with quarter-mile times often separated by mere tenths of a second.

The driving experience of these two cars, however, was markedly different. The Mustang, with its shorter wheelbase and lighter weight, offered a nimble, responsive feel that appealed to younger drivers and those who prioritized handling. The 442, built on GM's A-body platform, provided a more planted, substantial driving experience. It excelled in high-speed stability and long-distance comfort, characteristics that resonated with mature buyers who wanted performance without sacrificing refinement.

Culturally, the Mustang and the 442 occupied different spaces in the American automotive landscape. The Mustang became an instant icon, featured prominently in films, television shows, and popular music. It represented youth, freedom, and the American spirit of the 1960s. The 442, while not reaching the same level of pop culture saturation, cultivated a reputation as a "thinking man's muscle car." It appeared in car enthusiast magazines and was respected for its engineering prowess and balanced performance.

As the muscle car era progressed, both the Mustang and the 442 evolved to meet changing market demands and stricter emissions regulations. The Mustang underwent significant changes, including the controversial Mustang II of the mid-1970s, before returning to its performance roots. The 442, while maintaining its performance credentials, became increasingly luxurious, reflecting Oldsmobile's position as GM's near-luxury brand.

In retrospect, the 442 vs. Mustang comparison illustrates the diversity of the American performance car market during this golden era. The Mustang's success forced established brands like Oldsmobile to innovate and carve out their own performance niches. While the Mustang may have won the sales race and achieved greater cultural impact, the 442 proved that there was room in the market for a more sophisticated, engineered approach to high performance.

Today, both cars are highly sought after by collectors, each appreciated for its unique contributions to automotive history. The Mustang continues as a modern production car, carrying on its legacy of accessible performance. The 442, though no longer in production, remains a symbol of a time when Oldsmobile stood toe-to-toe with the best performance cars of the era, offering a compelling blend of power, comfort, and engineering excellence.

Section 12.5: 442 vs. Dodge Charger

The clash between the Oldsmobile 442 and the Dodge Charger represented more than just a battle of two muscle cars; it was a showdown between two distinct philosophies of American performance. While both vehicles embodied the spirit of the muscle car era, they approached it from different angles, each leaving an indelible mark on automotive history.

When it came to Mopar muscle, the Dodge Charger was a force to be reckoned with, and the 442 had its work cut out in matching Chrysler's offering. The Charger, introduced in 1966, quickly became an icon of the muscle car era with its sleek fastback design and powerful engine options. The 442, on the other hand, had been refining its formula since 1964, evolving from an option package to a standalone model that blended performance with a touch of luxury.

On the drag strip, these two titans often found themselves side by side, each vying for quarter-mile supremacy. The Charger, particularly in its R/T (Road/Track) configuration, was a straight-line monster. With engine options like the 440 Magnum and the legendary 426 Hemi, it could clock impressive quarter-mile times. The 442, not to be outdone, countered with its own array of powerful engines, including the formidable 455 cubic inch V8. While the Hemi-equipped Chargers often had the edge in raw acceleration, the 442 was no slouch, and its more balanced approach usually proved advantageous in real-world driving conditions.

Aesthetically, the 442 and the Charger represented two distinctive design languages. The Charger's flowing lines and hidden headlights gave it a look of barely contained aggression, while the 442 opted for a more refined, muscular appearance. The Charger's interior, with its full-length console and pistol-grip shifter, screamed performance. At the same time, the 442's cabin struck a balance between sportiness and comfort, reflecting Oldsmobile's position as GM's "innovative" division.

In terms of technology and innovation, both cars had their strengths. The Charger introduced features like an electric sunroof and a "Tuff" steering wheel, while the 442 boasted advanced suspension geometry and the innovative FE2 suspension package. Oldsmobile's focus on engineering refinement often gave the 442 an edge in handling and overall drivability, while the Charger's raw power and striking looks captured the imagination of many enthusiasts.

The racing heritage of both cars further cemented their performance credentials. The Charger made a name for itself in NASCAR, with its aerodynamic Daytona variant becoming a legend. The 442, while less prominent in circle track racing, proved its mettle in drag racing and road racing events. The success of both cars on the track translated to increased prestige and sales on the showroom floor.

Ultimately, the comparison between the Oldsmobile 442 and the Dodge Charger reveals two different approaches to the muscle car formula. The Charger embodied the "win on Sunday, sell on Monday" ethos with its racing success and bold styling. The 442, while no less capable, offered a more refined package that appealed to buyers seeking performance without compromising comfort and drivability.

Both cars left an indelible mark on the muscle car era, each with its loyal following and unique place in automotive lore. The 442 vs. Charger rivalry exemplifies the diversity and excitement of the muscle car era, a time when American automakers pushed the boundaries of performance and style, creating legends that continue to captivate enthusiasts to this day.

Section 12.6: The 442's Unique Selling Points

The Oldsmobile 442 stood out in the crowded muscle car market not just for its raw power, but for a combination of factors that made it truly unique. At the heart of the 442's appeal was its engineering excellence. Oldsmobile's engineers were known for their meticulous attention to detail, and this showed in every aspect of the 442. From its finely-tuned suspension to its precisely calibrated engine, the 442 was a marvel of automotive engineering that often outperformed its contemporaries in real-world driving conditions.

One of the 442's most distinctive features was its ability to balance luxury with performance. While many muscle cars of the era were stripped-down, bare-bones performance machines, the 442 offered a level of comfort and refinement that was typically reserved for more upscale vehicles. Plush interiors, advanced sound insulation, and smooth ride quality set the 442 apart from its rougher-edged competitors. This luxury-performance balance appealed to a more mature, sophisticated buyer who wanted the thrill of a muscle car without sacrificing comfort.

The 442 also boasted a range of exclusive features and options that weren't available on other muscle cars. The W-30 package, for instance, was a factory performance option that included a fiberglass hood with functional air scoops, a high-performance camshaft, and a special air induction system. These performance enhancements weren't just for show; they significantly boosted the car's power output and gave it a distinct edge on the street and strip.

Brand loyalty played a significant role in the 442's success. Oldsmobile had cultivated a loyal customer base over the years, known for valuing quality, innovation, and understated elegance. These customers saw the 442 as a natural progression of Oldsmobile's values. This car could deliver exhilarating performance without compromising on the refinement they had come to expect from the brand. This loyal following helped sustain the 442's popularity even as the muscle car market became increasingly competitive.

The influence of the 442 extended far beyond its production years. Its success inspired future GM performance models, setting a precedent for how to blend power with sophistication. The 442's approach to performance, emphasizing handling and overall driving experience over straight-line speed, foreshadowed the direction that many performance cars would take in subsequent decades.

Perhaps most importantly, the 442 represented Oldsmobile's commitment to innovation and pushing boundaries. It wasn't content to simply match its competitors; it aimed to surpass them in every way possible. This spirit of innovation became a hallmark of the 442, earning it a special place in the hearts of enthusiasts and cementing its status as more than just another muscle car. It was a testament to what American engineering could achieve when it set its sights on excellence.

Section 12.7: Legacy and Lasting Impact

The Oldsmobile 442's contribution to muscle car culture is nothing short of legendary. As we examine its lasting legacy, it becomes clear that this iconic vehicle has left an indelible mark on the history of the automotive industry. The 442 wasn't just a car; it was a symbol of American engineering prowess and a testament to the golden age of performance vehicles.

From its inception, the 442 played a crucial role in shaping muscle car culture. It challenged the notion that high performance was exclusive to specific brands, proving that Oldsmobile could compete with, and often surpass, its more widely recognized rivals. The 442's blend of power, handling, and luxury set a new standard for what a muscle car could be, influencing both its contemporaries and future generations of performance vehicles.

In the collector market, the 442 has shown remarkable resilience and appreciation. While some muscle cars have seen their values fluctuate wildly over the years, the 442 has maintained a steady upward trajectory. This trend speaks volumes about its enduring appeal and the respect it commands among enthusiasts. Scarce models, such as the W-30 package or convertible variants, have become highly sought-after prizes for serious collectors, often fetching premium prices at auctions.

The influence of the 442 extends far beyond its production years. Modern performance cars, even those outside the GM family, owe a debt to the engineering principles and design philosophy pioneered by the 442. The concept of a mid-size car with a big engine, enhanced suspension, and a luxurious interior continues to inform the development of today's sports sedans and muscle cars. Features that were innovative on the 442, such as hood scoops for ram-air induction and heavy-duty cooling systems, have become standard equipment on high-performance vehicles.

One of the most telling indicators of the 442's lasting impact is the vibrant enthusiast community that continues to celebrate this iconic model. Car clubs, online forums, and social media groups dedicated to the 442 are active and passionate, sharing restoration tips, organizing meetups, and keeping the spirit of the car alive. This ongoing enthusiasm ensures that the legacy of the 442 is passed down to new generations of car lovers.

In the annals of automotive history, the Oldsmobile 442 occupies a special place. It's remembered not just as a fast car, but as a well-rounded performer that could hold its own on the drag strip, carve through corners, and cruise in comfort. Automotive historians and journalists consistently rank the 442 among the top muscle cars of all time, praising its combination of power, handling, and sophistication.

The 442's legacy is also evident in its influence on Oldsmobile and GM's later performance models. The lessons learned from developing and marketing the 442 informed future projects, helping to shape vehicles like the Oldsmobile Cutlass 442 of the 1980s, and even influencing the performance divisions of other GM brands.

As we reflect on the legacy of the Oldsmobile 442, it's clear that its impact extends far beyond its impressive performance figures or sales numbers. It represents a high-water mark in American automotive design and engineering, a perfect storm of power, style, and innovation that continues to captivate car enthusiasts decades after the last model rolled off the assembly line. The 442 didn't just participate in the muscle car era; it helped define it, leaving a lasting legacy that ensures its place among the greatest American cars ever produced.

ABOUT THE AUTHOR

Todd Bandel is an accomplished author specializing in informational history books, with a particular focus on the automotive industry. Drawing from 40 years of experience as an automotive technician, Todd combines deep expertise and passion to enlighten readers about the historical nuances of automobiles. Todd currently resides in San Diego, California, where he continues to explore and write about his enduring interest in automotive history.

Mechanicaddicts.com

www.ingramcontent.com/pod-product-compliance
Lightning Source LLC
Chambersburg PA
CBHW020653220526
45464CB00001B/411